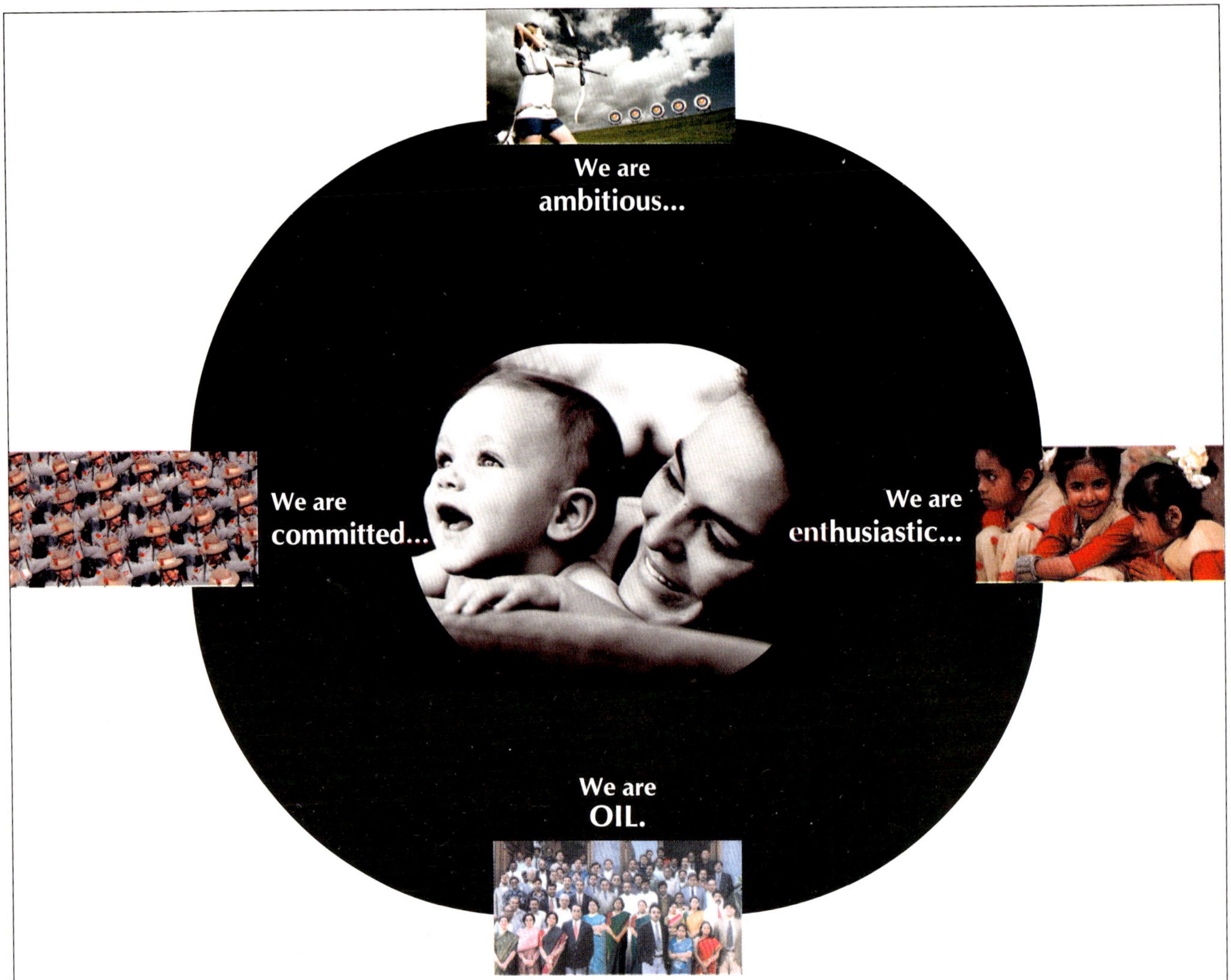
We are ambitious...
We are committed...
We are enthusiastic...
We are OIL.

When it comes to protecting the environment,
our commitment goes down-to-earth.

vedanta
Carrying on a tradition
that dates back to 500 BC.
Rajasthan, the land of world's oldest Zinc smelting producers. HZL – the proud inheritor of this unique age-old archaeometallurgical heritage.
Already 6th, HZL continues its journey to become 3rd largest Zinc producer in the world.
Hindustan Zinc Limited
http://www.vedantaresources.com
Making India proud.
Newfields

Cleaner is cheaper

Cleaner is cheaper

Case studies of corporate environmental excellence

Volume 5

Editor
R K Narang

The Energy and Resources Institute

ISBN 81-7993-118-8

Rs 1500/US $70

Published by
TERI Press
The Energy and Resources Institute
Darbari Seth Block
IHC Complex
Lodhi Road
New Delhi – 110 003
India

Tel. 2468 2100 or 4150 4900
Fax 2468 2144 or 2468 2145
India +91 • Delhi (0) 11
E-mail teripress@teri.res.in
Web www.teriin.org

Printed in India by Innovative Designers and Printers, New Delhi

Contents

Foreword

The issue of protecting the environment in a timely manner has been attracting the attention of policy-makers, civil society, and business for several decades now, particularly since Rachel Carson published her book *Silent Spring* in 1962. Up to that point, the developed world pursued the relentless goal of rapid growth without any regard to environmental consequences. As a result, the London smog of the 1950s, in which thousands of people died, and the case of large-scale inversion in the atmosphere in Donora, Pennsylvania, stand out as significant examples of the dangers from neglecting the environment.

The year 2007 marks the birth centenary of Rachel Carson, a pioneer and a fearless campaigner who used her knowledge of science to sensitize society in North America on the dangers of unmitigated use of chemicals and pesticides. This was clearly the start of a new era, because following her resolute and persistent advocacy of environmentally clean solutions, North America, followed by Europe, brought about major changes in the quality of water and air, and took in hand major afforestation programmes that corrected the imbalances of the past. By the 1990s, most of North America and Europe had created a clean environment for their citizens; but on the global front, there remains a different story to tell.

The IPCC (Intergovernmental Panel on Climate Change) has just completed three reports of the Working Groups dealing with the physical-science basis of climate change; impacts, adaptation, and vulnerability; and mitigation. These have created a major stir and have convinced leaders, as well as the public at large, on the cumulative damage created by industrialization over the last century and a half, which has resulted in an unprecedented increase in the concentration of GHGs (greenhouse gases) in the atmosphere leading to climate change and its various impacts, most of which are adverse in nature.

The Kyoto Protocol was designed to address the problem and laid down targets for reducing the emissions of GHGs, but it is now clear that not only were these targets inadequate, even though they represented an important first step, but the exclusion of some countries from the committed action was a major gap in what should have been a truly global effort. Solutions in the future will require government policies and the involvement of civil society, but it is business and industry that would have to really find the solutions that are essential and implement them with a sense of urgency. The impacts of climate change will leave no sector of the economy untouched, and are likely to affect prospects for business in a variety of ways. It would be proved over time that cleaner, in this context as well as on issues related to the local environment and natural resources, is cheaper, if we are to consider the costs of inaction and continuing with practices that are environmentally dirty.

This volume of *Cleaner is cheaper* deals, of course, with some outstanding case studies of companies that have shown responsibility and initiative in conceptualizing and implementing a series of measures in practice of the 'cleaner is cheaper' philosophy. However, it is only a matter of time before

Indian industry would also be required to mount actions for addressing global environmental problems, particularly the mitigation of GHGs to meet the threat of climate change. The philosophy of 'cleaner is cheaper' would then acquire a global relevance and dimension.

As it happens, actions for the mitigation of GHG emissions also have substantial co-benefits at the local level. Hence, the case studies described in this volume would provide a strong basis for exploiting some of these co-benefits for addressing both local environmental problems as well as the threat of climate change at the global level. These examples, therefore, have a larger and longer-term relevance than is apparent.

R K Pachauri
Director-General, TERI

Acknowledgements

Guidance
R K Pachauri
R K Narang
Annapurna Vancheswaran

Case study evaluators
Suneel Pandey
Harpreet Singh Kandra
Sumit Sharma
K V Santhosh Ragavan
Vishal Verma
Sachin Kumar
Sridhar Chidambaram
Kokil Gupta
Anshuman
Shubha Pandey

Research and writing
Jaya Kapur

Coordination
Jaya Kapur
V B Suneetha
Subrat Kumar Sahu

Copy-editing
Richa Sharma

Design and production
R Ajith Kumar
S Gopalakrishnan
R K Joshi
Tamal Basu
T Radhakrishnan

Marketing and outreach
Aarti Khosla
Meenakshi Mishra
Sandeep Sood

Introduction

With the Indian economy continuing on its high-growth trajectory, there is a general upbeat mood of confidence and expectation in business circles, both within the country and internationally. However, there also continues to be serious concern about some critical issues that have come to figure prominently on the national agenda. On the one hand, infrastructure bottlenecks in the power, transport, and communication sectors inhibit the fullest possible utilization of the opportunities thrown up by liberalization and globalization. On the other hand, the exclusion of large sections, particularly of the poor, from the benefits of development and the exponential degradation of our natural resource base cast doubts about the sustainability of the path that we have embarked upon.

It is important that the corporate sector, which is in the forefront of the country's economic resurgence, also takes due cognizance of these issues so that timely corrective steps and viable long-term solutions can be put in place. In order to successfully tackle the twin problems of poverty eradication and ecological restoration, partnerships between all stakeholders are clearly the need of the hour.

It is indeed reassuring to observe that awareness about the need for and a genuine commitment towards responsible action for integrating sustainability components into corporate strategy are clearly increasing amongst Indian industry. Not only have many companies set up environment management departments, but there is also a veritable scramble to obtain certification under international codes for HSE (health, safety, and environment) management. While this is essential for the export market, it is equally important in the overall national interest; for, without ensuring adequate standards in these aspects, India cannot lay claim to being a world-class player.

The threat of global climate change and its potential dangers are looming large and cannot be wished away, but has to be faced squarely and sensibly. It is also abundantly clear that rapid economic growth and development bring with them social costs by way of resource depletion and environmental fallout. In order to achieve sustainable levels of growth and development, these costs have to be ameliorated through well-thought-out policies and actions by government as well as business. At first sight, this may appear to be a daunting task that detracts from the immediate objective of increasing production, sales, and profit, but a closer examination would reveal that not only is it essential for survival in the long run, but also makes immensely good economic sense. The good news is that there is plenty of very useful R&D (research and development) around the world, including within India, to draw upon.

It is in this context that the fifth edition of *Cleaner is cheaper* seeks to highlight some of the outstanding innovative measures being taken by Indian industry for effective environmental management, including pollution control, water conservation, waste water utilization, energy efficiency, and the like. The focus on cleaner processes and systems, rather than end-of-pipe solutions, has come to be accepted as a more effective strategy, but still needs to be adopted and put into practice on a far wider scale. The fact that each of the initiatives has also yielded financial gains only goes to confirm the adage that 'it pays to be environment-friendly.' The examples documented in this book should motivate other organizations to benefit from these experiences so that they can be up-scaled and improved upon. It should also serve as a source of information to policy-makers, academics, and members of civil society at large.

The compendium contains a total of 25 case studies from some of the leaders of Indian industry, both in the public and private sectors. The case studies have been classified into the following four broad categories.

1 Environmental management system
2 Process improvement
3 Water/waste water management
4 Waste management and resource conservation

Many of the case studies, however, actually have cross-cutting themes and objectives and incorporate more than one of the above categories, thereby reflecting the integrative character of these efforts. Truly effective environmental management and sustainable resource utilization are a multi-disciplinary task.

With a turnover of more than Rs 860 billion, the annual production of the ONGC (Oil and Natural Gas Corporation Ltd) in 2005/06 included 24.4 MT (million tonnes) of crude oil and condensate, and 22.6 billion cubic metres of natural gas amounting to 84% of India's total oil and gas production. It is also the largest Indian multinational corporation with overseas investments of about $2.75 billion. All of the company's operational facilities and rigs – both onshore and offshore – maintain health, safety, and environmental management systems and were recommended for certification based upon ISO 9001, OHSAS 18001, and ISO 14001. ONGC is also the first public sector enterprise to have received DNA approval for the CDM (clean development mechanism) under the Kyoto Protocol. Further, the group has undertaken several projects under the CSR (corporate social responsibility) programme. ONGC generates large quantities of hazardous oily sludge and effluent soil waste, which was earlier tackled by storing the waste in special pits to prevent percolation and consequent contamination of ground water. This still did not fully stop seepage and overflow and also involved high construction costs. Using the bioremediation technology developed by TERI, the company now applies oilzapper that degrades the oil in sludge, soil, and water, and leaves a non-hazardous end-product, which can be used for greenbelt development. This treatment at the Lakwa site in Assam has yielded considerable cost savings.

Amongst the top 10 companies in India, NHPC (National Hydroelectric Power Corporation Ltd) has completed 12 hydel projects with an installed capacity of 4145 MW (megawatts), and has 12 more projects under construction with an installed capacity of 5332 MW.

Although hydro power projects use renewable and non-polluting resources for power generation, there are environmental impacts that always need attention. In recognition of this, NHPC, at its Uri power station in the Kashmir Valley, has various EMPs (environmental management plans) in place to address energy conservation, disposal of biomedical waste, installation of oil-catching devices, phasing out of CFC (chlorofluorocarbon) gas, and afforestation. A particularly significant initiative, in association with NEERI (National Environment Engineering and Research Institute), has been the innovative biotechnological approach for the restoration of 95 hectares of dumping sites with 5.5 million metres of excavated material.

Wipro Technologies Ltd is one of the world's leading technology service providers, with one of the largest R&D centres. Wipro has ISO certification through its initiatives towards energy and water conservation, waste management, and safety norms amongst which its 'green building' at the software development facility in Gurgaon, Haryana, is an example for others to emulate. In the construction of the 181 000 square feet of built-up area, environment-friendly materials like fly-ash-based aerated concrete blocks, recycled wood, ceramic tiles, and emission-free steel furniture have been utilized. Water efficiency and water conservation is achieved through recycling waste water, sprinkler and drip irrigation, and rainwater harvesting. A specially designed air-conditioning system combined with solar water-heating devices has resulted in a 30% reduction in energy consumption. All these measures have achieved significant savings in recurring costs with only a nominal increase in the initial capital cost. Wipro's Gurgaon Software Development Centre clearly demonstrates the considerable opportunities available in adopting 'green technologies' in construction and resource management to contribute towards good environmental practices while, at the same time, securing financial returns.

One of the largest manufacturers of steering systems for the automobile industry, Sona Koyo Systems Ltd, produced 730 000 steering systems with a turnover of over Rs 4000 million in 2005/06. As part of its CSR programme, the company runs a health-care institution and contributes towards upgrading schools in the vicinity of its manufacturing facility in Gurgaon, Haryana. It has also installed a state-of-the-art waste water treatment plant and rainwater harvesting systems.

An important initiative taken by Sona Koyo is involves the elimination of hazardous chromium toxicity commonly present in paint waste. By designing an efficient modified painting process and substituting the use of chromium in passivation with nitric acid, the company has been able to not only meet all quality requirements but also prevent groundwater pollution and occupational health hazards, not to mention the resultant annual cost saving of Rs 127 228.

These are merely indicative of a number of other best practices featured in this volume, which make a significant contribution to the growing literature in this important field.

Ashok Jaitly
Distinguished Fellow
TERI

WATER / WASTE WATER MANAGEMENT

1
Century Rayon, Century Textiles and Industries Ltd

2
Exide Industries Ltd

3
Hindustan Zinc Ltd

4
Indian Farmers Fertiliser Cooperative Ltd

CASE STUDY

1

Century Rayon, Century Textiles and Industries Ltd

RECOVERY OF WATER USED IN THE BACKWASHING OF SAND FILTERS IN THE WATER TREATMENT PLANT

Background

The Cenutry Spinning and Manufacturing Company Ltd set up its first cotton textile plant, called the Cotton Textile Mills, in October 1897. Today, Century Textiles and Industries Ltd is one of the largest composite textile mills in India, manufacturing a quarter of a million metres of cotton textile each day. It is also a leader in rayon and tyrecord, chemicals, cement, pulp and paper, and shipping.

The company's division at Shahad (Figure 1), District Thane, Maharashtra, commenced operations in 1956, with an initial capacity of five tonnes of VFY (viscose filament yarn) per day. Today, after successive capacity expansions, Century Rayon is the largest producer of VFY in India, and has a market-share of approximately 30%. The division's annual turnover for the period 2004/05 was in excess of Rs 4500 million.

Environmental profile

The company has an environment policy and environment cell at the facility level. The company has on-site and off-site emergency plan; all necessary equipment such as alarms, hydrant systems, fire extinguishers, and fire buckets are provided, and mock drills are periodically carried out.

Air pollution

Century Rayon takes effective measures to control and mitigate pollutants and wastes generated on the site. SO_2 (sulphur dioxide) is generated during the manufacture of sulphuric acid, from the coal-fired boiler, and DG (diesel generator) set. Double conversion-double absorption equipment is provided at both acid plants and outgoing SO_2 is continuously scrubbed with dilute caustic to minimize SO_2 emission (Figure 2). SPM (suspended particulate matter) is generated in the coal-fired boiler and is controlled by the ESP (electrostatic precipitator); adequate chimney height is provided for better dispersion in ambient air. NO_x (oxides of nitrogen) generated in the DG set and boiler house are well below the permissible limit. CS_2 (carbon di-sulphide) and H_2S (hydrogen sulphide) are liberated during the manufacture of VFY and tyre yarn. They are removed from spinning machines, from ducts connected to exhaust fans and chimney of adequate height to disperse the

Figure 1
Century Rayon, Shahad, District Thane, Maharashtra

Figure 2
SO_2 scrubber

gases in the atmosphere. Shutters have been provided in all the spinning machines to control dispersion of the gases into the work environment; the concentration of the gases is well within the stipulated limit in both the stacks, as well as in the work environment. H_2S generated in the manufacture of CS_2 is scrubbed with caustic solution to form sodium sulphide. A mist eliminator system has been provided in the sulphuric acid plant, and acid mist is well within the allowable limit.

Water pollution

The main water pollutant in rayon industries is zinc. Century Rayon houses a full-fledged ETP (effluent treatment plant) and zinc reduction system to control zinc and other parameters, namely, pH, BOD (biochemical oxygen demand), COD (chemical oxygen demand), oil and grease, and TSS (total suspended solids). The effluent flow of 11 000 m3 (cubic metres)/day is neutralized by the addition of lime solution in the equalization tank, and a hydaulic mixing system has been provided in the equalization tank to keep the solids in suspension. The equalized effluent is then pumped to the clarifoculator (Figure 3). Any overflow from the clarifoculator is led to the aeration tank, where fixed aerators have been provided, as shown in Figure 4.

Hazardous and industrial waste

Century Rayon is a member of the MWML (Mumbai Waste Management Ltd); sulphur and zinc waste are sent to their common waste disposal facility. The company has a centralized sulphur melter for the sulphuric acid and CS_2 plants. Sulphur waste generated from sulphur melters is collected by MWML and disposed off.

Industrial paper waste and waste oil are sold to recyclers registered with the MPCB (Maharashtra Pollution Control Board) and CPCB (Central Pollution Control Board). Cinder ash and fly ash are also given to outside parties for the manufacture of bricks. Other wastes are segregated at source and dumped at a designated site owned by the company.

Figure 3
Clarifloculator at the effluent treatment plant

Figure 4
Fixed aerators at the effluent treatment plant

Resource conservation

Energy and water conservation are given high priority at Century Rayon. Since 2003/04, the company has implemented 489 schemes for electric energy saving, and five schemes for thermal energy saving. The economizer-cum-superheater has been replaced with a new two-stage economizer where the water and gas circuits are counter-current to each other—the final water outlet goes to the boilers and the gas temperature going to the final absorption tower has reduced to 130 °C, implying greater heat recovery. Similarly, energy conservation projects were undertaken in the compressed air system and after-treatment drier.

The consumption of water in households and at the plant has been reduced each year through the introduction of recycle and reuse technologies and by rainwater harvesting. Total water consumption has thus reduced from 4.42 MGD (million gallons per day) in 1999 to 3.74 MGD in 2005.

Other initiatives

Environmental awareness is embedded into each of the employees, and is incorporated in their induction programmes. Environment Day is annually celebrated, and employees partake in poster and slogan competitions. Students from about 50 schools in the vicinity are also invited.

The company has a tree plantation programme through which over 23 000 saplings have been planted over the past 10 years. Tree plantation programmes are also arranged for different cooperative housing societies in the region, with the help of MPCB officials. Century Rayon carries out several CSR (corporate social responsibility) activities through its social welfare programme, health care and medical services progamme, and education and training.

Through the welfare programme, the company has set up two blocks of toilets in a neighbouring slum and provides aid during natural calamities like floods and earthquakes. It is working with the government and other voluntary institutions to implement the Government of India's 20-point programme. Century Rayon also contributes to the rural development programmes of two charitable trusts – Gramin Janaseva Sangh and Vasavdatta Foundation – that reach out to over 3000 persons each year. Through its medical programme, Century Rayon has set up a 66-bed hospital and eight-bed ICU (intensive care unit); arranges mobile dispensary and heath check-up camps; and regularly organizes blood donation, cancer detection, and family planning camps. The company also has a high school in English, Hindi, and Marathi medium and the Birla College of Arts, Science, and Commerce that has been conferred with a five-star rating by the NAAC (National Assessment and Accreditation Council). It has also set-up Cenray Mahila Pragati, a non-political and not-for-profit social organization for women to take part in social, cultural, and religious programmes. The organization also houses a Toddlers Academy for the preliminary education of children.

Case study

Before the implementation of its water conservation project at its WTP (water treatment plant), Century Rayon had 11 sand filters in its WTP, which were being used to filter water received from the clarifloculators. The backwashing of these filters was done on a timely basis as per the cleaning requirement to ensure that the sand filters did not get choked. Hence, 10 backwashes were carried out each day and filter backwash water was being drained and wasted. As a result, the company lost about 0.2 MGD of water. Table 1 shows backwash water consumption from April 2003 to March 2004.

An analysis was carried out, after which it was decided to collect backwash water and recycle it back to the 8 MGD clarifloculator, where the water would be treated along with raw water from the river, rechlorinated to kill bacteria, and then refiltered. As the working of softeners in the water plant was stopped, two brine tanks of 60 m^3 capacity each were utilized for the collection of sand filter backwash water instead. The brine tank was modified slightly, giving a necessary slope at the bottom and interconnecting the tanks at the bottom; necessary outlets

TABLE 1
BACKWASH WATER CONSUMPTION (APRIL 2003–MARCH 2004)

Month	*Year*	*Average water consumption for backwashing sand filters (MGD)*
April	2003	0.260
May	2003	0.254
June	2003	0.256
July	2003	0.171
August	2003	0.177
September	2003	0.174
October	2003	0.192
November	2003	0.218
December	2003	0.272
January	2004	0.280
February	2004	0.237
March	2004	0.263
Average		0.2295

MGD – million gallons per day

and drain nipples were provided to act as 'settlers'. The existing drain channels were then diverted to the brine tanks by providing civil diversion and necessary piping. High/low-level instrumentation in the brine tanks was provided, to enable it to automatically operate the backwash recycle pump as per high/low level of the tanks. Drain valves in the tanks were provided with timers to drain the sludge from the bottom every hour for one-and-a-half minutes. The modified system can be seen in Figure 5, which depicts the sand filters both before as well as after implementation of the project.

After the installation of the project, the filter backwash of about 910 m^3/day, which was earlier being drained, is now recovered and recycled back to the raw water storage tank along with river water, and is sent, once again, to the clarifloculator with the raw water for treatment purposes. As 10 backwashes were carried out each day before the implementation of the project, now, approximately 910 000 m^3/day of water (that is, 0.2 MGD), which was previously being drained back to

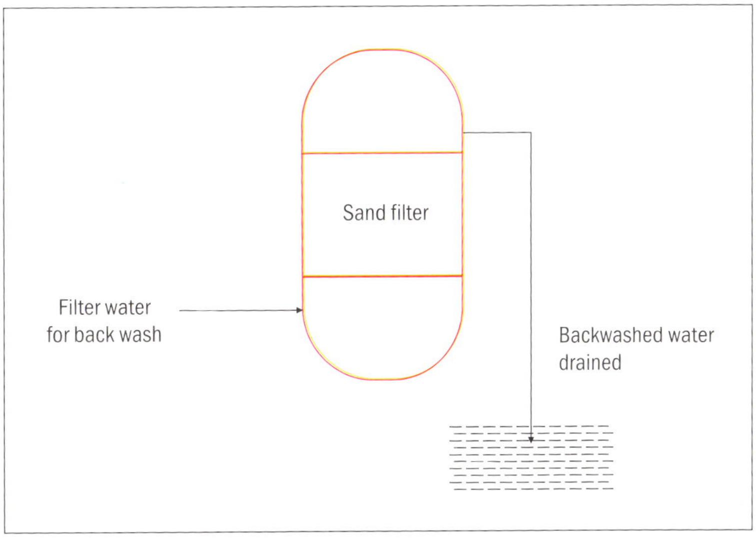

Figure 5a
Before implementation of the water conservation project

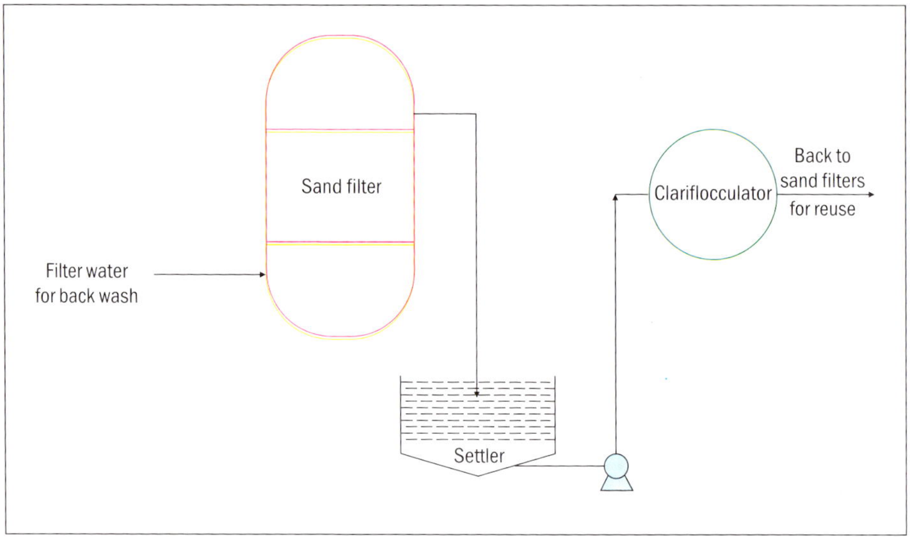

Figure 5b
After implementation of the water conservation project

the river, is being reused. This reuse of water also greatly lowers the use of fresh water. Moreover, as Century Rayon was paying the government Rs 8.59 for each m^3 of raw water drawn from the Ulhas river in 2004/05 (in addition to Rs 1.10/m^3 spent in pumping power and filtering the raw water), the company saves approximately Rs 3.2 million per annum.

CASE STUDY at a glance

- Century Rayon, Century Textiles and Industries Ltd, located at Shahad, Maharashtra, is the largest producer of VFY (viscose filament yarn) in India, with an annual turnover in excess of Rs 4500 million for the period 2004/05.
- Before implementation of the case study, the company's WTP (water treatment plant) had 11 sand filters, for which 10 backwashes were carried out each day to ensure that they did not get choked. This resulted in a loss of about 0.2 MGD of water.
- A modified system was designed so that backwash water could be collected and recycled back to the 8-MGD clarifloculator, where the water would be treated along with raw water from the river, rechlorinated to kill bacteria, and then refiltered.
- After installation of the modified system, filter backwash of about 910 m^3/day, which was earlier being drained, could now be recovered and recycled back to the raw water storage tank along with river water, and sent to the clarifloculator with the raw water for treatment purposes.
- The project has not only greatly lowered the use of fresh water, but has also resulted in a cost saving of Rs 3.2 million per annum.

CASE STUDY

2

Exide Industries Ltd

INTEGRATED APPROACH FOR POLLUTION PREVENTION THROUGH EFFLUENT MINIMIZATION

Background

Exide Industries Ltd (Figure 1) set up its Hosur *Taluk* Unit in Tamil Nadu's Krishnagiri district in 1997. The unit manufactures lead acid storage batteries for both industrial and automotive applications. Annual production in 2005/06 was 3 364 800 units, while the annual turnover for the same period was over Rs 4900 million.

Environmental policy

Exide Industries Ltd has an environment policy and an environment cell at both corporate and facility levels. The company has an EMS (environment management system), and risk assessment and disaster management plans are in place.

The facility takes several steps to control pollutants. Lead dust is a potential air pollutant released during the battery-manufacturing operation, for which adequate control equipment like mist scrubbers, bag filters, and electrostatic filters are provided. Periodic analysis of stack emissions, ambient air quality, and work zone quality is carried out. Adequate stack height is provided to control SO_2 (sulphur dioxide), and there are suction hoods, fume scrubbers, and blowers for sulphuric acid mist generated during the process.

Exide is a 'zero-effluent-discharge factory', and all the effluents generated are treated and reused. Sulphuric acid is one of the main water pollutants generated from the battery-manufacturing process; the process effluent generated is reduced from the consented quantity of 219 kL (kilolitres)/day due to the modified processes of recycling and reuse; waste water from machine cleaning and floor washing is used for

Figure 1
Exide Industries Ltd, Hosur *Taluk* Unit

gardening after treatment; reject water from the RO (reverse osmosis) plant, which is installed to reduce the feed water TDS (total dissolved solids) of DM (demineralized) plant, is used for toilet flushing and floor washing; ion-exchange-based system for water softening for cooling application is eliminated and instead RO-treated water for cooling is used to avoid the generation of high TDS effluent; waste water from domestic application is collected and treated separately using the aerobic method; and so on.

Lead scrap is the major hazardous waste generated at Exide; the entire amount is sent to MoEF (Ministry of Environment and Forests)-authorized smelters for recovery. Waste generated from the effluent treatment operation is disposed off in secured landfills. Waste oil and spent oil generated both from the manufacturing operation and the canteen are collected and sent to authorized recyclers. Oil-soaked gloves and cotton waste are incinerated. Garden and canteen wastes are converted to bio-compost.

Resource conservation is a priority area for Exide. Water conservation is ensured through the recycling of water and tertiary treatment of effluent. Further, the unit has a full-fledged rainwater harvesting system with a storage capacity of 31 600 kL to ensure that no rain water runs out of the factory premises. Emphasis is also laid on energy conservation, as energy costs are sizable for all lead acid battery manufacturers.

CSR (corporate social responsibility) at Exide revolves around five major areas: education, health, environment, women empowerment, and relief and assistance. Table 1 highlights some of the key CSR initiatives taken by the company.

A feasibility study for a project under the CDM (clean development mechanism), in association with IISc (Indian Institute of Science)–Bangalore, on the utilization of ETP (effluent treatment plant) sludge is under way.

Case study

Repeated monsoon failures in the Hosur region pushed Exide Industries to carry out careful analyses of its activities, so that appropriate water conservation measures could be taken. Studies on the site showed that approximately 60% of the total water intake was going either to the STP (sewage treatment plant) or ETP. Exide identified the reduction of effluent generation to be an effective way by which it could conserve water. Prior to implementation of the case study, effluent generation at the unit was 210 kL/day.

The case study was initiated by introducing visual control systems at various effluent-generation points. The effluent characteristics of each stream were communicated to the concerned department in-charge, mainly to study the variation and to take corrective action. A six-stage method was developed entailing: (1) monitoring and measurement, (2) segregation, (3) process modifications, (4) recycling and reuse, (5) volume reduction, and (6) reduction and training.

Monitoring and measurement

In order to prepare the water balance, flow metres were installed in all the main freshwater and effluent lines. Total generation per day was meticulously measured and an electronic reporting system was introduced. Effluent generation data was reviewed on a daily basis.

TABLE 1
CORPORATE SOCIAL RESPONSIBILITY INITIATIVES AT THE HOSUR FACTORY

Area	*Interaction with*	*Activities undertaken in the community*
Education and training	School authorities, government bodies, and, educational institutions	▪ Sponsoring an underprivileged child at a local school ▪ Providing books and uniforms at local panchayat schools ▪ Environment awareness at panchayat schools ▪ Vocational trainees for skill development ▪ Project work guidance for graduate and diploma engineers ▪ Factory visits by students from various institutes
Health and welfare	Panchayats, NGOs, local community, and primary health officials	▪ General and specialty health camps ▪ Blood donation camps ▪ Pulse polio immunization ▪ Remuneration to health educators ▪ Ambulance services to accident victims and pregnant women ▪ Refreshments and transport for blood donors ▪ Banners, transport, and volunteers during health camps ▪ Development of infrastructure at Bagalur primary health centre ▪ Provision of furniture for training hall to facilitate training of health care workers
Environment	BDO, panchayats, factory inspector, fire officer, sub collector	▪ 10 km motorable road from main road to Chichuraganapalli village ▪ Bus shelter at Bilapur village ▪ Funds for safe drinking water to Hadhenalli panchayat ▪ Funds for drainage, street lights, and temple renovation ▪ ISO 14001 certification ensuring good environmental performance ▪ 12.5 hectares of green belt development ▪ Rainwater harvesting of 30 000 kL in 2005 ▪ No rain water and treated effluent let out of the factory premises ▪ 50% reduction in effluent generation ▪ 22% reduction in raw water consumption ▪ Periodic testing of air, water, and soil samples
Women's empowerment	Local vendors	▪ Face mask and hand gloves being procured from physically handicapped and women self-help groups ▪ Assistance to self-help women group at Vista Ashram, Bidadi
Assistance and relief	NGOs and local organizations	▪ Contributions to Chief Minister's tsunami relief fund and earthquake relief fund ▪ Appliances to disabled during World Disabled Day, poor feeding in the local community, etc. ▪ Development of sports ground at neighbourhood school

BDO – block development officer; NGO – non-governmental organization; kL – kilolitre

Segregation

Analysis of individual streams revealed that effluent generated in one area could be reused in another area as a raw material for another unit process. Segregating the streams allowed the effluent characteristics and cost of treatment to various departments to be studied.

Process modifications

The process modifications include switching from tank formation to jar formation, which greatly enables the reduction of effluent generation. The entire automotive battery manufacturing is jar-formed, and 90% of the unit's total production exploits this concept, leading not only to a significant reduction in effluent generation, but also to a reduction in the DM water requirement and a tangible drop in electric power consumption.

Recycling and reuse

Recycling using RO technology

Exide, Hosur, is the first battery-manufacturing industry in India to utilize RO technology for the tertiary treatment of effluent. Permeate from the RO plant is used for machine cooling, whereas reject with high TDS is evaporated. Salt from the solar evaporation plan is disposed off in a scientifically secured landfill.

Acid recovery system in formation

Sulphuric acid is generated during the charging of battery plates in the plate formation operation. Dilute solution of sulphuric acid is used as the electrolyte in the formation tanks, in which the plates are arranged as cathodes and anodes. External electrical potential is applied to the series of cathodes and anodes to carry out the charging operation.

Electrolytic reaction taking place during the charging operation results in the formation of sulphuric acid. Thus, the concentration of the acid, which is initially four per cent, increases up to eight per cent. In order to maintain the acid concentration at four per cent, approximately 40 litres of acid used to be drained from each tank, and the reduced volume was made up by the addition of DM water into each jar. Acid removal was carried out after every charging cycle of approximately 20 hours.

Acid removal was done manually, and the acid removed from the tanks was taken to the ETP, where it was neutralized using alkali. Exide was incurring high operating costs in the ETP, and large quantity of sludge was being generated as a result of this process.

To overcome these problems, a mechanized acid collection system (Figures 2a and 2b) was designed at the purification unit to purify the acid for reuse and recycle. The pumping system designed consisted of trolley-mounted double diaphragm pumps in place of manual siphoning to take out the excess acid. This acid is transferred to underground storage tanks in pipelines to avoid contamination. Acid from the underground collection tank is then pumped through a basket filter, micron filter, activated charcoal filter, and specially designed heavy metal removal filter for the removal of iron and other heavy metal ions. Treated acid is stored in overhead storage tanks, and from there it is used for acid dilution and for jar formation of automotive batteries.

DM water recovery system

Analysis on waste water generated from the washing of negative plates showed that sulphuric acid was the only contaminant, and that its concentration level was below 300 PPM (parts per million). With the help

Figure 2a
Acid collection system

Figure 2b
Forming area

of an external agency, Exide, Hosur, was able to purify wash water at the ion-exchange-based sulphate removal unit. Cation and anion resin units were also installed for the purification process as a small quantity of metal ions was also found in the wash water.

Recycling the water used for machine cleaning

Waste water generated from the cleaning of paste mixing and pasting machines contain lead oxide paste. Overflow from the wash water sedimentation tank is collected and filtered using a micron filter and is reused.

Acid from the field failed batteries

The acid drained from the cut-open analysis of field-failed batteries is collected, filtered, and reused for acid dilution. Figure 3 shows the acid collection for field-failed batteries.

Volume reduction

During the charging of batteries, a small quantity of acid mist escapes through the acid-filling valves, even though the fume extraction system consists of a suction hood, wet scrubber, blower, and the required height of stack is provided. Some quantity of mist condenses and falls into the water bath for battery cooling. This was making the cooling water acidic, and hence, the water had to be changed frequently. Hence, Exide, Hosur designed a mist eliminator whereby the acid mist gets condensed and falls back inside the battery. This technology minimizes acid loss during charging and prevents the water from becoming acidic.

Figure 3
Acid collection system for field-failed batteries

Education and training

Following the belief that the key to achieving environmental targets is through providing skills, competence, and knowledge to its employees, Exide, Hosur, has kept aside six person-days per year for each employee to spend on learning.

Benefits

The case study has resulted in a reduction of 45 300 kL/annum of effluent, reduction in raw water consumption of 45 300 kL/annum, reduction in solid waste generation of 300 tonnes/annum, and energy conservation of 58 500 kWh (kilowatt-hours) per annum. Monetary savings as a result of the case study can be seen in Table 2.

TABLE 2
COSTS AND SAVINGS OF SYSTEMS

Description of the system	*Cost of the system* (Rs)	*Cost saving per annum* (Rs)
Acid recovery system for forming acid	800 000	4 179 000
DM water recovery unit for formed plate wash water	800 000	50 000
Collection and filtration system for cut opened battery	60 000	10 000
Filtration system for machine cleaned water	40 000	8 000
RO system for high TDS water from the regeneration of DM plant and solar evaporation	2 500 000	–
Conversion of through battery washing system to circulation type	1 000	4 590
Mist eliminators for batteries while charging	500 000	37 500
Tank formation to jar formation	22 000 000	2 600 000

DM – demineralized; RO – reverse osmosis; TDS – total dissolved solid

CASE STUDY

at a glance

- Exide Industries Ltd's Hosur *Taluk* Unit in Tamil Nadu's Krishnagiri district manufactures lead acid storage batteries for both industrial and automotive applications.
- Annual production in 2005/06 was 3 364 800 units while the annual turnover for the same period was over Rs 4900 million.
- Repeated monsoon failures led Exide to look into ways by which it could conserve water; the reduction of effluent generation was found to be an effective way to achieve this.
- The company devised a six-step methodology entailing: (1) monitoring and measurement, (2) segregation, (3) process modifications, (4) recycling and reuse, (5) volume reduction, and (6) education and training.
- The case study has resulted in a reduction of 45 300 kL (kilolitres)/annum of effluent, reduction in raw water consumption of 45 300 kL/annum, reduction in solid waste generation of 300 tonnes/annum, energy conservation of 58 500 kWh (kilowatt-hours) per annum, as well as monetary savings.

CASE STUDY

3

Hindustan Zinc Ltd

Water conservation measures taken for reduction of freshwater consumption in beneficiation plant

Background

HZL (Hindustan Zinc Ltd) is India's largest producer of zinc. It is part of the London-listed Vedanta Resources Plc, a metals and mining major with aluminium, copper, and zinc operations in India, Australia, and the United Kingdom.

HZL's Rampura Agucha Mine produced 720 050 MT (million tonnes) of zinc concentrate and 65 194 MT of lead concentrate in 2005/06. Its annual turnover for the same period was Rs 6.22 billion, while that of HZL was Rs 38.77 billion.

Environmental profile

Rampura Agucha Mine has an EMS (environment management system) in place since the commissioning of the mine, and consists of regular monitoring of stack emissions, ambient air quality measurement, and work zone and water quality, etc. The mine has been certified for ISO 14001: 1996 and 2004, and OHSAS 18001: 1999.

The mine was awarded by the Ministry of Power, Government of India for power conservation in the mining sector. For three consecutive years from 2003/04 to 2005/06, Rampura Agucha Mine was awarded for overall excellence in the Ajmer region during the Mines Environment and Mineral Conservation week celebrations, held under the aegis of the Indian Bureau of Mines. Recently, Rampura Agucha Mine was selected for the Gem Granite Environment Award by FIMI (Federation of Indian Mineral Industry).

The major air pollutants emitted by mining operations at Rampura Agucha Mine include SPM (suspended particulate matter), RSPM (respirable suspended particulate matter), and noise. The air quality management system at the mine is in place to minimize the generation of air pollutants at the site. Track mounted DTH (down-the-hole) drills with dust extraction systems are used, and all drilling machines, as shown in Figure 1, have wet drilling systems. Haul road dust is suppressed through regular water sprinkling. A 28-kL (kilolitre) water sprinkler is shown in Figure 2. Dust from ore crushing and handling equipment is being controlled by a dust extraction system through wet scrubber. Fine ore is stored in buns to minimize dust generation, and venture scrubbers are installed for dust removal at the primary, secondary, and tertiary crushers and fine ore bins. All heavy earth-moving equipment have been provided with exhausts, which are located away from the road surface and discharge upwards to prevent dust. The installation of an SAG (semi-autogenous grinding) mill has eliminated secondary and tertiary crushing, hence eliminating dust generation at source in dry crushing. Moreover, a 50–100-metre wide green belt has been developed around the mining complex to capture fugitive emissions of dust. A total of 272 000 trees have been planted in 2005/06. Moreover, monitoring of air quality is carried out by qualified personnel using state-of-the-art equipment, ensuring that emissions are within limits.

Figure 1
Drill machine with wet system

Figure 2
Water sprinkler to suppress dust on haul roads

In order to control water pollution and achieve zero-discharge status, the water quality of the mine pit and wells in and around the mine complex is monitored on a weekly and monthly basis, respectively. The tailing of the beneficiation plant is treated with hydrated lime prior to pumping to the tailing dam, and the reclaimed water is pumped back to the plant for reuse. A reclaim water reservoir of 50 000 m^3 (cubic metres) capacity was constructed to reduce water evaporation losses at the tailing dam. Water from all sources is pumped to this reservoir for recycling.

Major categories of hazardous waste include tailings, used/spent oil, and discarded containers. Tailings are stored in tailing dam after neutralization.

Efforts are being made to recover the metals from the tailings by bio-leaching. Used/spent oil is collected in MS (mild steel) drums and stored in a shed until sold to recyclers registered with the MoEF (Ministry of Environment and Forests)/CPCB (Central Pollution Control Board). Empty sodium cyanide drums are decontaminated with sodium hypochlorite solution, making them non-hazardous.

Steps are taken to reduce power consumption at the beneficiation plant and diesel consumption in HEMM (heavy earth-moving machinery), topsoil management, spread environmental awareness, maximize water conservation, etc. The facility has environmental risk management and disaster management plans as well. The Emergency Preparedness Response Plan was prepared and implemented at the time of ISO 14001 and OHSAS 18001 implementation.

Case study

Expansion and the need for water conservation

The Rampura Agucaha Mine is situated in a water-scarce area with semi-arid climate. There is no source of water available near the mine premises, and water is drawn through a 60-km long pipeline from the radial well situated in the Banas river bed. During the inception of the mine in 1991, the production capacity was 3000 MT of ore per day. By 1999, the capacity of the mine was expanded from 3000 TPD (tonnes per day) to 4500 TPD. Efforts were started in 2003/04 for further expansion capacity of the mine to 11 400 TPD ore production and treatment. The major constraint for expansion was water availability, as the process of beneficiation is water intensive. To overcome this constraint, a number of water conservation initiatives were taken, such as a drive for water harvesting and recycling measures, which resulted in the reduction of freshwater consumption from 1.36 m^3/MT in 2002/03 to 0.66 m^3/MT in 2005/06, as detailed in Figure 3.

Measures towards water conservation

In the mine area, drains around the mine pit and waste dump are provided for the collection of run-off water, which is collected in a lined storage pond with an arrangement to pump it to the tailing dam. Lined drains have been provided at the tailing dam to collect and recycle any seepage. Water in the tailing pond is reused in the process plant; hence, zero-discharge status is maintained. Evaporation retardant chemicals are used at the tailing dam. There is a lined pond (Figure 4) of 5000 m^3 capacity for collecting and recycling any accidental spillages and

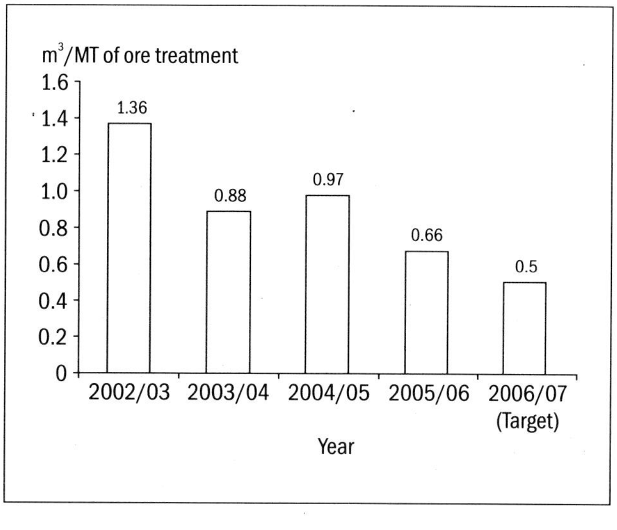

Figure 3
Water consumption (m^3/MT of ore treatment)

Figure 4
Lined pond

overflows. Level controllers are installed at this seepage pond to prevent the overflow of water and to maintain optimum level in the sump in order to allow continuous pumping to the reclaim reservoir, shown in Figure 5. The reclaim water reservoir of 50 000 m^3 was constructed to reclaim water from the tailing dam, mine, and other areas; reduce evaporation losses at the tailing dam; and recycle water to beneficiation plant without any treatment. An oil and grease trap was constructed for workshop effluent. As seen in Figures 6a and 6b, high-rate thickeners are used to improve water recovery from tailing and concentrates.

Moreover, domestic sewage water is reutilized for plantation after treatment. The use of flocculent in tailing thickeners for water recovery and use of reclaim water for the de-dusting system are further steps to conserve water. Reclaim water and turbo drill for auto-dilution replace fresh water in the flocculent dozing system. Mine pit water is pumped directly to the new water reservoir tank for recycling, and spillage water and HEMM washing water are pumped to the new reclaim water

Figure 5
Reclaim water reservoir

6a

6b

Figure 6
Tailing thickener

reservoir. In order to prevent the wastage of fresh water, level controllers have been installed in the pressing water tank. Strict vigilance and control to prevent/minimize spillages and overflows in the industrial area are maintained.

Future plans for further water conservation

In addition to successfully implementing the above-mentioned water conservation measures, the Rampura Agucha Mine has plans to reduce water consumption upto 0.5 m^3/MT of ore treatment 2006/07. This target is to be achieved through the introduction of further measures towards water conservation including setting up an STP (sewage treatment plant) of 425-kld (kilolitres per day) capacity in the colony area to reduce water consumed in horticulture works and another STP of 300-kld capacity in the mine area to be used in the process plant, thereby reducing freshwater consumption. Other initiatives planned include the control of pressing water in a pressure filter, use of recycled water instead of fresh water for cleaning the mill, transportation of tailing dam water under gravity to the reclaim reservoir, and use of wetting agent on haul roads on a trial basis.

Environmental and cost benefits

As a result of the measures undertaken at Rampur Agucha Mine, not only did the water consumption reduce from 1.36 m^3/MT in 2002/03 to 0.66 m^3/MT in 2005/06, but the plant now operates at zero discharge, and thus, no groundwater pollution is reported as assessed by several studies conducted at the site. Energy is conserved through the retrofitting of all thickeners, as the under-flow density increased from 25% (from conventional thickeners) to 45%, thus reducing the quantity of water pumped to the tailing dam and, in the process, preventing seepage.

Further, the case presents monetary savings of Rs 31.5 million per annum to the organization. Without implementing this project to conserve water, the Rampur Agucha Mine would have been unable to expand its operations, which may have resulted in substantial financial losses for the company as well as for the national economy in terms of valuable foreign exchange due to the import of zinc, lead metals, and concentrates.

CASE STUDY at a glance

- The Rampur Agucha Mine is a unit of the Rs 38.77 billion Hindustan Zinc Ltd.
- The mine produces zinc and lead concentrates, and achieved an annual turnover of Rs 6.22 billion in 2005/06.
- As the mine was increasing its operational scale, it became imperative for the plant to adopt practices to conserve water, without which further expansion would have been impossible in light of water scarcity.
- Water-saving measures included provision of drains around the mine pit and waste dump for collection of run-off water, which is collected in a lined storage pond; creation of a lined pond of 5000-m^3 capacity to collect and recycle accidental spillages and overflows; and reuse of all water in the tailing pond by the process plant.
- The reutilization of domestic sewage water for plantation after treatment, construction of oil and grease trap for workshop effluent, use of flocculent in tailing thickeners, and use of reclaim water for the de-dusting system were some other steps taken towards saving and recovering water resources.
- As a result of the various measures implemented at the mine, water consumption was reduced from 1.36 m^3/MT in 2002/03 to 0.66 m^3/MT in 2005/06.
- The plant is also operating at zero-discharge conditions.
- A monetary saving of Rs 31.5 million per annum was achieved by the organization.

CASE STUDY

4

Indian Farmers Fertiliser Cooperative Ltd

USE OF WASTE WATER FOR BIOLOGICAL NITROGEN FIXATION IN THE SOIL

Background

IFFCO (Indian Farmers Fertiliser Cooperative Ltd) has four fertilizer plants—the Kalok and Kandla units in Gujarat, and the Phulpur and Aonla units in Uttar Pradesh. IFFCO's Phulpur unit (Figure 1) produces about 1.7 MT (million tonnes) of urea annually. It achieved a turnover of over Rs 24 billion in 2005/06.

Environmental profile

All four of IFFCO's manufacturing units have implemented EMS (environment management system) and have been accredited with ISO 14001 certifications. Environment policies and departments exist at both corporate and facility levels. The facility has environmental risk and disaster management plans, and strictly complies with all the rules under the Environment Protection Act (1986). Recent awards received by IFFCO include the Best Pollution Control Implementation Award (2005) from Public Sector Today, Hyderabad, and The National Award for Excellence in Water Management (2005), from CII (Confederation of Indian Industries), Hyderabad.

Various steps are taken to ensure that pollutants produced on site are kept to their minimum. SPM (suspended particulate matter) and RSPM (respirable suspended particulate matter) are two of the air pollutants generated; ESPs (electrostatic precipitators) are provided in the flue gas paths, and a fuel switch-over from LSHS (low sulphur heavy stock)/FO (furnace oil) to natural gas in a boiler has been carried out to keep the

Figure 1
IFFCO, Phulpur

pollutant level low. Cyclone dust collectors have been provided to reduce coal dust and urea dust in the ambient air, and two natural draft prill towers of ample height have been provided to reduce the ground-level concentration of urea dust. Flue gas dispersion at a high altitude (100–200 metres) also reduces the ground-level concentration of SO_2 (sulphur dioxide) and NO_x (oxides of nitrogen). High-efficiency burners with combustion control systems have been provided in both the ammonia and steam generation plants for proper and efficient combustion of fuels. Fume scrubbers, dust suppression systems, and automatic sensors for the detection of ammonia, chlorine, hydrogen, etc. are some systems in place for combating fugitive emission.

The major organic waste in the effluent is urea, which raises the oxygen demand of effluent water. Hence, the urea plants have waste water treatment sections with a deep urea hydrolyser, which recovers CO_2 (carbon dioxide) and NH_3 (ammonia) from the waste water and recycles them back into the process. Approximately 2200 m^3 (cubic metres)/day of hydrolysed water is reused as boiler feed water in the steam generation plant after polishing in the ion exchange condensate polisher unit. BOD (biochemical oxygen demand) is also found in domestic sewage water for which a well-designed sewage-treatment-cum-recycle plant has been installed; treated water is used as cooling tower make-up and for irrigation. Acidic and alkaline effluents are collected in the neutralization pit, where they are neutralized and fed to the RO (reverse osmosis) plant for treatment, after which they are reused. Effluent containing oil and grease is passed through gravity oil separators or disc oil separators. Two guard ponds of 7000 m^3 and 5000 m^3 capacities have been provided for treating TSS (total suspended solids), and sufficient holding time is provided for the settling of the suspended particles. An ETP (effluent treatment plant) based on RO technology operates to treat industrial effluent; TDS (total dissolved solids) removal takes place in the RO plant. Hence, the entire treated effluent (approximately 5260 m^3/day) is recycled or reused.

The major categories of hazardous waste generated include spent catalysts and waste oil. The former are stored securely in closed metallic drums and are sold for metal recovery to parties authorized by the SPCB (state pollution control board). Waste oil is collected or reclaimed for gravity oil separators and mixed with fuel oil and pumped to steam generation plant, where it is used as a supporting fuel along with coal. Other wastes generated include ash from the steam generation plant, which is disposed off to cement industries, and lime sludge that is sold to outside parties and is also used as binding material in fly ash brick plants.

IFFCO Phulpur has implemented various programmes to lessen its environmental footprint. It has a number of schemes, for instance, to reduce CO_2 emission—they include energy-saving scheme in its ammonia plants to reduce fossil fuel consumption and steam requirement, and fuel switch-overs from heavy hydrocarbons such as LSHS/FO and naphtha to natural gas. Similarly, steps have been taken to reduce SO_2 emission. Other initiatives include rainwater harvesting and recharge, use of vermicomposting for conversion of biodegradable waste into manure, adoption of non-chromate cooling water systems in process plants, installation of chlorinators in cooling towers, utilization of ash from coal-fired boilers, etc.

IFFCO's initiatives extend to the wider farming community as well through its 158 FSCs (farmers service centres) across 10 states.

Balanced fertilization programmes, farmers' meetings, and crop seminars are carried out as part of these initiatives. Under CORDET (Cooperative Rural Development Trust), promoted by the company, IFFCO provides practical training to farmers to improve their skills in agricultural production, horticulture, dairy, poultry, bee-keeping, fruit and vegetable preservation, professional leadership, computer use, etc. IFFCO adopts villages for socio-economic development activities. Fodder is provided in drought-prone areas; medical check-ups (both veterinary and human) and medicine distribution are carried out; drinking water facilities are provided; assistance is given to school children, and literature containing information on farming practices for various crops, balanced use of fertilizer, soil testing, etc. is made available.

The company has prepared the following four projects under the CDM (clean development mechanism).

1 Reduction in steam consumption through revamping of ammonia plants
2 Feed switch-over from naphtha to natural gas in the ammonia plants
3 Installation of CDR (carbon dioxide recovery) plants
4 Fuel switch-over from naphtha/LSHS to natural gas in the ammonia plants

Case study

Introduction

The nutrient needs of growing plants can be met through a number of sources—mineral fertilizers, organic manures, recycled waste and its by-products, biological nitrogen, natural minerals, and, to a lesser extent, nutrients recycled through irrigation water and precipitation. These supplement the soil nutrient reserve for nourishing the crops.

Present-day intensive agriculture requires a high input of nitrogenous fertilizers, which is met by the production of ammonia from gaseous nitrogen. The cost of producing nitrogenous fertilizers has been escalating over the years, and it is clear that nitrogen produced in the industries cannot meet the total nitrogen needs of the world. Thus, the nitrogen needs of agriculture will have to be met from fertilizers, organic manures, efficient recycling, and nitrogen-fixing microorganisms.

Microorganisms are important components of soil; they complete their lifecycles in the soil, and hence, bring about a number of changes in physical, chemical, and biological properties of the soil. Therefore, these microorganisms are used to restore the fertility status of the cropland and field, and are used as important tools in wasteland reclamation as they can increase the productivity of soil.

The study

IFFCO carried out a detailed field study with the following objectives.

- To study the variation of biological nitrogen-fixing microorganisms in soil irrigated with factory effluent water, effluent + sewage water, sewage water, and bore well water during different seasons.
- To study the diversity patterns of nitrogen-fixing microbial population with variable irrigation during different seasons, at two different depths of soil profile.
- To study the effect of physico-chemical properties of soil on the nitrogen-fixing microbial population during different experimental conditions.
- To study the effect of minerals of soil on the nitrogen-fixing microbial population during experimental field conditions.

The study was conducted in 2002/03, on agricultural land situated near IFFCO's fertilizer factory at Phulpur *tehsil*, approximately 35 kilometres from the city of Allahabad. In the field experiment study, the comparative occurrence of four different nitrogen-fixing bacteria – namely, Rhizobium, Azotobacter, Azospirillum, and blue-green algae – have been recorded in three seasons – summer, rainy, and winter – using different types of irrigation waters – namely, treated effluent water from the nitrogenous fertilizer plant, treated effluent + treated domestic sewage water, treated domestic sewage water, and bore well water – at two different depths of soil (7.5 cm and 15 cm) under field conditions. Along with the individual population of these nitrogen-fixing microorganisms, the pysico-chemical parameters of the soil like pH, EC (electrical conductivity), organic carbon, available N (nitrogen), available P (phosphorous), K (potassium), Ca (calcium), Mg (magnesium), Fe (iron), Zn (zinc), Na (sodium), Cu (copper), and Mn (manganese) were also determined.

Results

Table 1 summarizes some of the results derived from the study. It shows the values of the highest and lowest number of different nitrogen-fixing microorganisms up to 15 cm of depth in the soil irrigated with different types of water during different seasons. It shows, for instance, that the Rhizobium population in soil at a depth of 7.5 cm varies from 1.5 × 104 to 1.3 × 105 cells per gram of soil. It also shows that the population is highest during the rainy season, followed by the winter, and then the summer season, irrespective of the type of irrigation water used. Table 2 provides a summary of the highest and lowest values of physico-chemical parameters like pH, EC, organic carbon, N, P, and K of soil irrigated with different types of water during different seasons, up to 15-cm depth, during the three different seasons.

TABLE 1
Summarized values of highest and lowest nitrogen-fixing microbial population with different types of irrigated water systems, during various seasons, and at different soil depths

Types of nitrogen fixing microorganisms (1000 CFU/gm of soil)	Soil depth (cm)	*Summer season* Highest value		*Rainy season* Highest value		*Winter season* Highest value	
		Highest value	*Lowest value*	*Highest value*	*Lowest value*	*Highest value*	*Lowest value*
Rhizobium	7.5	35 (3)	15 (1)	130 (2)	30 (4)	75 (2)	18 (1)
	15	32 (3)	8 (1)	170 (1)	17 (4)	60 (3)	30 (2)
Azotobacter	7.5	30 (3)	8 (1)	78 (4)	25 (3)	50 (3)	24 (4)
	15	38 (3)	8 (2)	58 (4)	16 (1)	60 (2)	15 (1)
Azospirillum	7.5	1.5 (3)	0.5 (2)	18 (1)	9 (4)	18 (1)	13 (3)
	15	1 (4)	0.4 (1)	12.5 (1)	2 (4)	17.5 (1)	8 (3)
Blue-green algae	7.5	8.8 (4)	4.8 (1)	9 (4)	1.5 (2)	7 (2)	2.5 (1)
	15	10.2 (3)	3.9 (1)	8.5 (4)	2 (2)	6 (2)	3.2 (1)

CFU – colony forming unit; (1) – effluent water irrigated soil; (2) – effluent + sewage water irrigated soil; (3) – sewage water irrigated soil; (4) – bore well water irrigated soil

TABLE 2
SUMMARIZED VALUES OF HIGHEST AND LOWEST PHYSICO-CHEMICAL PARAMETERS WITH DIFFERENT TYPES OF IRRIGATED WATER SYSTEMS, DURING VARIOUS SEASONS, AND AT DIFFERENT SOIL DEPTHS

Physico-chemical parameters of soil	Soil depth (cm)	*Summer season*		*Rainy season*		*Winter season*	
		Highest value	*Lowest value*	*Highest value*	*Lowest value*	*Highest value*	*Lowest value*
pH value	7.5	8.8 (4)	7.8 (1)	9.2 (3)	8.5 (2)	7.8 (3)	7.6 (1)
	15	9 (4)	8.1 (1)	8.8 (4)	8.6 (1)	7.9 (4)	7.6 (1)
EC value (dSm^{-1})	7.5	0.6 (1)	0.18 (4)	0.5 (3)	0.36 (2)	1.1 (2)	0.42 (4)
	15	0.5 (1)	0.16 (4)	0.62 (2)	0.25 (4)	0.8 (2)	0.37 (4)
Organic carbon (%)	7.5	0.6 (1)	0.25 (4)	0.5 (1)	0.26 (4)	0.76 (1)	0.4 (4)
	15	0.33 (1)	0.18 (4)	0.5 (3)	0.22 (4)	0.45 (2)	0.37 (1)
Available N_2 (kg/ha)	7.5	40 (1)	19 (4)	70 (1)	40 (4)	140 (1)	60 (4)
	15	25 (2)	23 (4)	80 (3)	45 (2)	180 (3)	50 (2)
Available P (kg/ha)	7.5	31 (4)	18 (3)	38 (1)	14 (2)	22 (4)	2 (2)
	15	28.5 (4)	16 (3)	28 (1)	10 (2)	0.5 (4)	4 (1)
Available K (kg/ha)	7.5	500 (1)	320 (3)	650 (2)	520 (4)	400 (1)	280 (4)
	15	400 (1)	300 (4)	810 (2)	600 (4)	500 (3)	280 (4)

EC - electrical conductivity; dSm^{-1} - deco-siemens per metre; kg/ha - kilograms per hectare; N - nitrogen; P - phosphorus; K - potassium; (1) - effluent water irrigated soil; (2) - effluent + sewage water irrigated soil; (3) - sewage water irrigated soil; (4) - bore well water irrigated soil

The summary of the highest and lowest values of minerals – Ca, Mg, Na, Cu, Mn, Zn, and Fe – present in Phulpur soil up to 15 cm of depth, being irrigated with different water systems during the summer, rainy, and winter seasons can be seen in Table 3.

Conclusion

Through the study, it was found that in an alkaline tract, like the one in which nitrogen-fixing microbial studies were conducted, pH, EC, organic carbon, and available nutrients are the parameters likely to affect the total number of a specific group of nitrogen-fixing microorganisms. The nature of irrigation water is likely to change pH, EC, and available nutrients in general. At the same time, however, temperature, which is governed by the seasons, has a marked effect, because even the organic matter content may change and the soluble salts may accumulate in the upper 7.5 cm of the soil.

The results clearly show that Phulpur soil gives shelter to all the four types of nitrogen-fixing microorganisms. The coexistence of all four nitrogen-fixing microorganisms in experimental plots is perhaps due to the suitable irrigation waters. The presence of essential mineral nutrients and the micronutrient in the treated effluent water + domestic sewage

TABLE 3

Summarized values of highest and lowest soil minerals with different types of irrigated water systems, during various seasons, and at different soil depths

Soil minerals and heavy metals (PPM)	Soil depth (cm)	Summer season Highest value	Summer season Lowest value	Rainy season Highest value	Rainy season Lowest value	Winter season Highest value	Winter season Lowest value
Available calcium	7.5	650 (1)	290 (4)	370 (1)	280 (4)	300 (3)	280 (2)
	15	600 (1)	330 (3)	390 (1)	300 (2)	360 (1)	305 (4)
Available magnesium	7.5	120 (1)	58 (3)	38 (1)	33 (2)	39 (1)	35 (4)
	15	115 (1)	52 (3)	37.5 (2)	36 (4)	40 (1)	37 (3)
Available sodium	7.5	500 (1)	125 (2)	400 (1)	130 (4)	210 (1)	100 (4)
	15	330 (1)	130 (4)	450 (1)	95 (2)	180 (1)	100 (3)
Available copper	7.5	1.3 (1)	0.38 (2)	1.2 (1)	0.65 (4)	0.85 (4)	0.6 (1)
	15	1.5 (3)	0.39 (4)	0.9 (3)	0.6 (4)	1.1 (4)	0.55 (1)
Available manganese	7.5	3.3 (3)	0.4 (4)	2 (3)	1 (4)	2.5 (4)	0.8 (1)
	15	1.6 (1)	1 (3)	2.2 (3)	1.5 (4)	2.2 (3)	1.2 (2)
Available zinc	7.5	1.22 (1)	0.6 (4)	1.5 (4)	0.65 (3)	1.65 (3)	0.7 (2)
	15	0.82 (1)	0.4 (4)	1.3 (4)	0.6 (3)	1.1 (3)	0.5 (1)
Available iron	7.5	5.3 (1)	1.8 (4)	6.5 (2)	1.8 (4)	0.8 (3)	5.8 (1)
	15	4.7 (3)	1.3 (4)	6 (4)	2 (3)	6 (3)	1.6 (1)

PPM – parts per million; (1) – effluent water irrigated soil; (2) – effluent + sewage water irrigated soil; (3) – sewage water irrigated soil; (4) – bore well water irrigated soil

from the IFFCO Phulpur fertilizer factory shows that there has been no soil pollution. If the composition of irrigation water remains at the present level, as has been used in the experimental plots, there is no possibility of soil deterioration, if microbial population is an index of soil health and its fertility. Hence, the increase of all four types of nitrogen organisms, with the use of treated effluent water and treated domestic sewage water, is a good sign for maintaining a high level of nitrogen in the soil.

The study shows that of the four types of nitrogen-fixing microorganisms, Rhizobium population is the highest in treated factory effluent-sewage water. Hence the study encourages farmers to use sewage and effluent water from the IFFCO Phulpur fertilizer factory.

CASE STUDY

at a glance

- IFFCO (Indian Farmers Fertiliser Cooperative Ltd) fertilizer plant at Phulpur, Uttar Pradesh, produces about 1.7 MT (million tonnes) of urea annually, and achieved a turnover of Rs 24 billion in 2005/06.
- IFFCO Phulpur conducted a study on the use of waste water for biological nitrogen fixation in the soil.
- The study was conducted in 2002/03, on agricultural land situated near IFFCO's fertilizer factory.
- In the field experiment study, the comparative occurrence of four different nitrogen-fixing bacteria – Rhizobium, Azotobacter, Azospirillum, and blue-green algae – were recorded in three seasons – summer, rainy, and winter – using different types of irrigation waters – treated effluent water from nitrogenous fertilizer plant, treated effluent + treated domestic sewage water, treated domestic sewage water, and bore well water – at two different depths of soil (7.5 cm and 15 cm) under field conditions. Physico-chemical parameters of the soil like pH, EC, organic carbon, available nitrogen, calcium, magnesium, iron, zinc, etc. were also determined.
- The study showed that Phulpur soil gave shelter to all the four types of nitrogen-fixing microorganisms, and that the coexistence of all four nitrogen-fixing microorganisms was perhaps due to the suitable irrigation waters. Hence, the treated effluent water + domestic sewage water from the IFFCO Phulpur fertilizer factory in irrigation proved harmless for nitrogen-fixing microorganisms, which improve and maintain soil fertility.
- The study also showed that of the four types of nitrogen-fixing microorganisms, Rhizobium micro population was highest. This along with the irrigation with treated factory effluent + sewage water provides a favourable condition for the cultivation of leguminous crops, as opposed to paddy and wheat.

WASTE MANAGEMENT AND RESOURCE CONSERVATION

5
Grasim Industries Ltd

6
Oil and Natural Gas Corporation Ltd, Lakwa, Assam

7
Reliance Industries Ltd, Hazira

8
Reliance Industries Ltd, Patalganga

9
Tata Sponge Iron Ltd

10
UAL–Bengal, UAL Industries Ltd

CASE STUDY

5

Grasim Industries Ltd

RESOURCE RECOVERY AND RECYCLING

Background

Grasim Industries Ltd was established in 1954 and is a flagship company of the Aditya Birla Group. It is the largest producer of VSF (viscose staple fibre) in the world, with a market-share of about 21%. Grasim is also India's largest producer of sodium sulphate, a by-product of VSF that is widely used in the paper and pulp, detergent, glass, and textiles industries.

The major products manufactured at Grasim's staple fibre division (Figure 1), based in Nagda, Madhya Pradesh, are VSF, sodium sulphate, CS_2 (carbon disulphide), sulphuric acid, and captive power. The Nagda division has an installed capacity of 151 475 tonnes of VSF and an employee strength of over 3000. It achieved a turnover of Rs 12.53 billion in the financial year 2005/06.

Environmental profile

Grasim has a certified EMS (environment management system); it is ISO 9002, ISO 14001, OHSAS 18001, and SA 8000 certified. The company has an environment policy, an environment cell responsible for EMS, and a safety department responsible for occupational health and the safety management system. There is a separate corporate SHE (safety, health, and environment) cell that coordinates, initiates, and monitors SHE initiatives across the group.

Grasim has a number of air pollution control systems. There are 11 CS_2 recovery system for spinning machines; a Klaus kiln for the CS_2 plant; and water condensers, brine condensers, and an oil scrubbing system for the recovery of CS_2. Four gas-scrubbing systems for tail gases with 50 metre tall chimneys and mist eliminators are provided at the acid plants to abate SO_2 emission. To control SPM (suspended particulate matter), the auxiliary section is provided with cyclones and water scrubbers, the CS_2 plant with a dust extraction-cum-century scrubbing system, and power plants with ESPs (electrostatic precipitators) and ash handling plants. Ear-muffs and ear-plugs are used to keep noise pollution to the minimum. The encasing of blowers and provision of silencers in vents also control noise pollution.

Grasim's water pollution facilities include neutralization tanks and a circular equalization tank for pH generated; grit chambers, a biological

Figure 1
Staple fibre division, Nagda, Madhya Pradesh

reactor with 22 aerators, sludge thickener system, sludge dewatering system, etc. for suspended solids, BOD (biochemical oxygen demand), and COD (chemical oxygen demand) generated; and oxidation ditches for domestic effluent.

The plant also takes hazardous and industrial waste management measures. Used oil is sold to an authorized recycler, while ETP (effluent treatment plant) sludge is dewatered, dried, mixed with coal, and then fed to coal-fired boilers for heat recovery. Hard mass, mainly containing charcoal and a negligible amount of sulphur, is fed to coal-fired boilers for energy recovery, and waste sulphur from the acid plant, as and when generated, is transferred to the sulphur recovery pit for filtration and reuse. Over 95% of fly ash, an industrial waste, is utilized in the manufacture of bricks and cement; the remaining fly ash is stored at an authorized site as per the MoEF (Ministry of Environment and Forests) guidelines. Piece-up waste generated during the regeneration of cellulose is washed, neutralized, and collected in tractor trolleys and transported to an authorized site for development. Resource conservation initiatives taken by Grasim are outlined in the case study.

Grasim is involved in the socio-economic development of its industrial complex. Wasteland development, horticultural activities in collaboration with the state forest department, medical and public health services, provision of pure drinking water facilities to villages, and provision of education for villagers are some examples of activities undertaken by Grasim. The company has also taken considerable steps in the development of a green belt around its plant, and over 225 000 trees have been planted on 188 hectares.

Grasim has environmental risk management and disaster management plans in place. The company has received several awards for environment and safety, some of which include the Safety Innovation Award by the Institute of Engineers (2006), the Greentech Gold Industrial Safety Award (2005), the CII (Confederation of Indian Industries) Industrial Safety, Health, and Environment Management Award (2004), and the Stockholdm Industry Award (2004).

Case study

Grasim takes several initiatives towards the conservation of resources through recovery and recycling. The two guiding principles for this entailed, first, identification of different streams that were getting drained from various sections, resulting in losses in the form of thermal energy, chemicals, or water; and second, finding applications on a scientific basis for use at the nearest section to reduce the requirements of fresh resources. Grasim not only takes initiatives to control consumption of input, but also takes steps towards generating 'wealth from waste'—several examples of both are furnished below.

Initiatives to control the consumption of inputs

Recovery of sodium sulphate and sulphuric acid

Sodium sulphate is a by-product of VSF manufacturing and is obtained in dissolved form. Its recovery requires the evaporation of excess water and crystallization of concentrated solution. The recovery process has been improved with the use of multi-stage flash evaporators that are highly economical, as opposed to double- and triple-effect evaporators.

The use of cluster-type spinnerets in the system permits lower acid levels on account of improved spin bath availability to the innermost filaments, hence reducing carry-over losses as compared to the flat-type spinnerets. Scientifically designed systems are utilized for tow washing and squeezing, thereby further reducing carry-over losses of the spin bath. Water containing sodium sulphate and sulphuric acid, which used to transport the cut fibre to the next stage, is recycled, resulting in higher concentration of the chemicals and is part recovered in the multi-stage flash evaporators. Savings, as a result, amount to approximately 18 m^3/hour of water, 900 kg/day of sodium sulphate, and 180 kg/day of sulphuric acid, in addition to heat recovery.

Recycling of viscose and viscose-containing wash water

Continuous filters have been installed in place of filter presses in various stages of viscose filtration to avoid significant losses of viscose during filter press cleaning. These filters have not only reduced the loss of viscose, but have also reduced the manpower requirement. The filter press is thus only used to filter reject viscose—the recovered viscose and wash water containing cellulose and caustic are once again used in the process after filtration.

Wash water generated due to the washing of Simplex, after completion of every batch of Xanthation reaction, which was earlier drained, is not being mixed with product resulting in saving of water, cellulose, and caustic. Moreover, all viscose tanks and filters require scheduled cleaning, for which draining of leftover viscose is essential. It is thus collected in movable trolleys or trays and pumped back to the receiving tanks using movable transfer pumps.

Water recycling in cooling towers

Cooling towers consume a substantial amount of water in the unit. With several cooling towers of different capacities serving different applications, guidelines were prepared for reducing water consumption, including the following.

- Effective control of drift losses by using efficient drift eliminators
- Increasing the cycles of concentration through the use of appropriate chemical treatment
- Reducing cooling tower blowdown by the use of side stream filtration
- Use of treated effluent at the rate of 400 m^3/hour for make-up in cooling towers

Recycling in the water treatment plant

Water used for the back washing of sand filters, softeners, ion exchangers, and the final rinse of the mixed bed is recycled, saving 4925 m^3/day of water. The weak solution of caustic soda, left after regeneration of ion exchangers, is used in the cooling towers of the main plant to neutralize the acidity in circulating water. Moreover, waste lean solution of brine from the chemical division is used to regenerate the water softener, resulting in the reuse of 10 TPD (tonnes per day) of NaCl (sodium chloride) and a cost saving of Rs 800 000, in addition to curbing pollution generated at the site.

Recycling of condensate with the use of available heat

Condensate from the multi-stage flash evaporators, generated at the rate of 100 m^3/hour, is used for fibre washing, spin bath filter back washing, and spin bath tank cleaning. Steam condensate from fibre dryers, at

17 m^3/hour, is sent back to the water treatment plant, after recovering heat from it, and is used as feed for the water demineralization plant. Condensate from the salt dryers, at 2 m^3/hour, is used for fibre washing, after recovering heat from it. Finally, steam condensate from the CS_2 plant, at 8 m^3/hour, is collected in a common tank and pumped back to the acid plant for use as boiler feed water, resulting in a saving of both water and energy.

Generating wealth from waste

Burning of ETP sludge for energy generation

On the successful implementation of zinc-free process technology, the sludge from the ETP has become non-hazardous and is thus burnt in the existing coal-fired boilers for energy recovery after drying in the steam dryers. This has completely eliminated the sludge disposal problem in addition to heat recovery, equivalent to 8 TPD of coal.

Biogas generation from dairy effluent

The BOD/COD-rich streams of dairy effluent are treated in an anaerobic filter treatment plant with generation of biogas. The approximate reduction in BOD/COD by 80%–90% is achieved.

Utilization of fly ash

With ongoing efforts, 100% of fly ash generated at the captive thermal power plants is utilized for clay-brick manufacturing. A fly ash–cement brick-manufacturing unit with a capacity of 30 000 bricks per day has been installed on the premises. This unit further allowed Grasim to meet the total requirement of bricks on the industrial complex and no clay bricks were procured after September 2003. The progressive utilization of fly ash at the company can be seen in Table 1. Fly ash also has other uses, including road construction, construction of hollow blocks, for mixing in small quantity with RCC (reinforced cement concrete) for the construction of buildings (giving added strength to the structure by reducing air voids), and for utilization in cement manufacturing.

Sulphur recovery from H_2S-laden tail gases

The generation of H_2S (hydrogen sulphide) is unavoidable in the CS_2 manufacturing process. This, along with uncondensed CS_2, is converted to elemental sulphur through the Klaus process. The H_2S- and CS_2-bearing gases are partially burnt in controlled quantity of air and passed through a bed of bauxite catalyst to get elemental sulphur.

TABLE 1
PROGRESSIVE FLY ASH UTILIZATION

Financial year	*Generation* (MT)	*Utilized* (MT)	*Utilized* (%)	*Remarks*
2000/01	212 640	171 333	81	Remaining fly
2001/02	234 086	200 229	86	ash was utilized
2002/03	284 616	241 312	85	for landfilling
2003/04	224 913	194 273	86	purposes
2004/05	258 677	237 924	92	

MT – million tonnes

Benefits

The maximization of resource utilization through resource recovery and recycling has reduced the generation of waste at the source itself. The recycling of chemicals and other inputs such as acid and water has resulted in enhanced production, as well as cost savings, as shown in Figure 2. The consequent reduction in influent load has increased the effectiveness of effluent treatment and helps in rendering final discharge or emissions harmless to the environment. Further, the generation of value-added products such as energy from ETP sludge, bricks from fly ash, and biogas from dairy effluent are additional advantages that can be termed 'creating wealth from waste.'

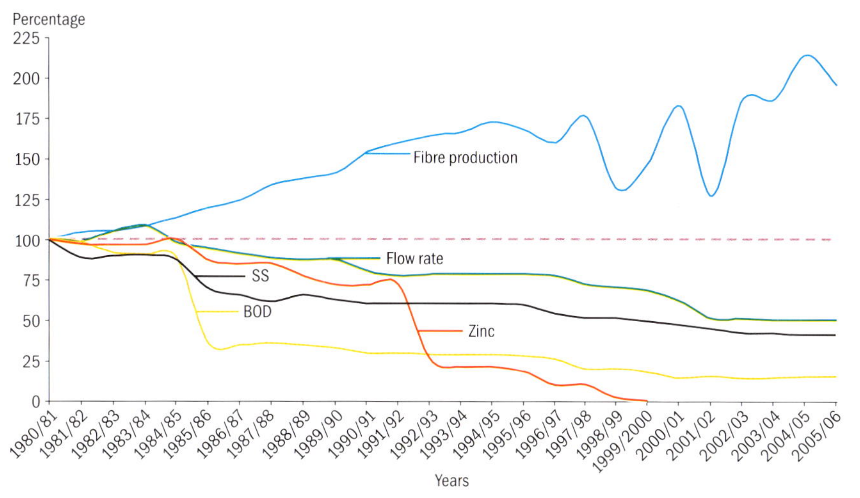

Figure 2
Progressive reduction in pollution load

CASE STUDY

at a glance

- Grasim Industries Ltd, a flagship company of the Aditya Birla Group, is the largest producer of VSF (viscose staple fibre) in the world.
- Efforts towards recovery and recycling at Grasim include not only curbing consumption of inputs, but also 'generating wealth from waste'.
- Various measures were taken for the recovery of sodium sulphate and sulphuric acid; through the use of cluster-type spinnerets, scientifically designed systems for tow washing and squeezing, and recycling water containing sodium sulphate and sulphuric acid that used to transport the cut fibre to the next stage, Grasim saves approximately 18 m^3 (cubic metres)/hour of water, 900 kg/day of sodium sulphate, and 180 kg/day of sulphuric acid, in addition to heat recovery.
- Other steps towards conservation include recycling of water in the cooling tower, recycling in the water treatment plant, recycling of viscose and viscose-containing wash water, and recycling of condensate with the use of available heat.
- In converting waste into wealth, Grasim uses ETP (effluent treatment plant) sludge for energy generation; creates biogas from dairy effluent, reducing 80%–90% of BOD (biochemical oxygen demand)/COD (chemical oxygen demand); and sulphur recovery from H_2S (hydrogen sulphide) – laden tail gases.
- Another significant initiative is the utilization of fly ash for brick-making. The company's fly ash-cement brick-manufacturing unit with a capacity of 30 000 bricks per day meets the total requirement of bricks on the industrial complex.
- These initiatives have reduced the generation of waste, curbed pollutants generated on the site, conserved valuable resources, increased production, and resulted in large cost savings for the company.

CASE STUDY

6

Oil and Natural Gas Corporation Ltd

Bioremediation of oily sludge at ONGC

Background

ONGC (Oil and Natural Gas Corporation) is one of the largest players in the oil and natural gas industry in the country. The ONGC Group – comprising ONGC, wholly-owned subsidiary OVL (ONGC Videsh Ltd), and MRLP (Mangalore Refinery and Petrochemicals Ltd) – produced 1.11 million barrels of oil per day in 2005/06. The ONGC Group turnover for the same period was Rs 864.14 billion.

Environmental profile

All of ONGC's operational facilities and rigs – both offshore and onshore – are certified for HSE (health, safety, and environment) management under relevant international and national codes. All facilities maintain an EMS (environment management system), and as on 31 March 2005, over 98% of ONGC operations were recommended for certification on quality, occupational health safety, and EMS based on ISO 9001, OHSAS 18001, and ISO 14001.

ONGC was India's first public sector enterprise to receive DNA (Designated National Authority) approval from the Government of India for CDM (clean development mechanism) under the Kyoto Protocol. The six CDM projects undertaken include gas flaring reduction projects pertaining to different ONGC Assets in Mumbai High, Neelam and Heera, Ankleshwar, Ahemdabad, Mehsana, and Cauvery Asset. Another three CDM projects are in the pipeline.

The company takes several measures to ensure that air and water pollutants, as well as hazardous and industrial wastes, are kept at minimum levels. Air pollutants are generated mainly from technical gas flaring and from flue gases from the DG (diesel generator) sets and cogeneration plants. Monitors are installed and ambient air quality monitoring vans are used. Air monitoring is also carried out by third parties approved by the respective SPCBs (state pollution control boards). Flare gas recovery is carried out at many of the installations, and a heat recovery system to heat the therm oil and crude oil by way of heat exchange from flue gases is available at many installations. Acoustic enclosures are available in the cogeneration plant to contain sound pollution.

ONGC has 23 ETPs (effluent treatment plants) with a total treatment capacity of 70 000 m^3 (cubic metres)/day. PWCs (produced water conditioners) with in-built oil recovery systems are installed at all offshore process platforms. Hydro-cyclones for improved recovery of oil will soon be used at the PWCs. For the treatment and recycling of waste water in the process of onland drilling operations, mobile ETPs are being used. Waste pit systems are used at all onland drill sites to recycle waste water.

Chemical sludge and tank bottom sludge are the major solid hazardous wastes generated by ONGC. Tank bottom sludge is sold to approved vendors, while chemical sludge is collected in lagoons with leachate facilities where water is drained to reduce the quantity of sludge. As discussed in the case study, bioremediation techniques are used to treat the sludge/effluent. Drill cuttings are washed with water, checked, and used for landfilling; spent oil and slop oil are recycled or reused; used lead acid batteries are sold to authorized vendors for recycling; and biomedical waste is incinerated.

ONGC has taken up other environmental projects and community development activities as well. In a major project carried out in collaboration with the Forest Development Corporation of Maharashtra, large-scale mangrove plantations have been developed near Uran, Raigad, where about 26 700 saplings have been planted over 7.5 hectares of land. A second similar project, carried out in collaboration with the BNHS (Bombay Natural History Society), entailed the plantation of approximately 500 000 mangrove saplings, R&D (research and development) on various species, documentation of species, training and awareness programmes, etc.

CSR (corporate social responsibility) activities are carried out under the SEDPs (socio-economic development programmes) wherein schemes for the promotion of education; health care; community development; calamity relief; and sports, arts, and culture are developed and implemented.

Case study

Oily sludge and oil-contaminated soil

ONGC generates large quantities of tank bottom oily sludge and oily effluent soaked soil waste, which constitute a major challenge for hazardous waste and environment management. Further, oil spills also take place in the company's day-to-day operations. Prior to implementation of the project, ONGC has been tackling these wastes by storing oily sludge (hazardous hydrocarbon waste) and oil-contaminated soil in synthetic line pits or specially designed and constructed pits, with a leachate collection system and polymer lining to prevent the percolation of contaminants into groundwater. After each pit was filled, another was constructed in order to accommodate the waste. This approach to manage oily sludge and oil-contaminated soil presented many problems such as limited seepage, high costs of constructing seepage-free pits, and overflow during the rainy seasons, which, in turn, was affecting surrounding farming properties, etc. Moreover, ONGC found that this method of transporting oily waste from the generation point to the disposal site was akin to transferring the problem from one point to another; to the environment pollution was still taking place.

Oilzapper as a solution

Various challenges and difficulties arising out of ONGC's traditional method led it to seek other alternatives. The company approached TERI, New Delhi, that has developed a new method, using bioremediation for the degradation of oily sludge, waste water effluent sludge, and other oily waste, including formation water. TERI's method entails the application of oilzapper (Figure 1), a bacterial system consisting of a consortium of different types of microorganisms that degrade the oil in contaminated systems such as sludge, soil, and water. Oilzapper biodegrades aliphatic, aromatic, nitrogen, sulphur, oxygen-containing compounds, and asphaltene fractions of crude oil or oily sludge. Upon applying oilzapper to oily sludge, waste water effluent sludge, and oil-contaminated soil or water, microbes degrade the toxic hydrocarbon compounds of the sludge;

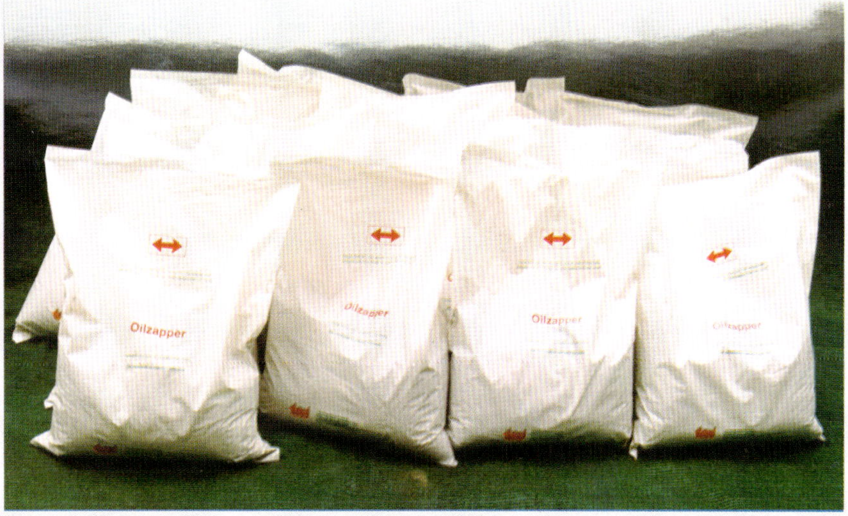

Figure 1
Oilzapper developed by TERI

what remain as a result are non-hazardous end products or metabolites that are harmless to the environment.

The bioremediation technique not only resolved ONGC's problem of oil contamination and environment pollution, but also saved considerable land that was earlier being used for the construction of pits. Further, oilzapper also proved to be environment-friendly and cost-effective.

Lakwa, Assam project

A project was undertaken at an ETP (effluent treatment plant) sludge pit at ONGC Lakwa, Assam Asset, for the bioremediation of 966 tonnes of oily sludge. The area covered by the site was 80 metres × 15 metres. The site was prepared by first clearing it of stones, grass, debris, etc. Ten random samples of oily sludge were collected in polythene bags and divided into two parts—one set of samples was transported to TERI's Microbial Biotechnology Laboratory, New Delhi, for analysis of the related parameters, while the other was sent for parallel analysis at ONGC's laboratory. Then, a calculated quantity of oilzapper culture, containing live sludge-degrading bacteria, was applied to the ETP sludge at the bioremediation site, as seen in Figure 2. A nutrient mixture was also sprayed on the oily sludge. Once applied, the mixing of oily sludge and oilzapper bacteria was done by tilling the site either manually or by using a tractor.

Results at the Lakwa, Assam site

Four bore wells were installed to monitor groundwater quality at the bioremediation site. Water samples were collected from the bore wells—one set of samples was used to analyse pH, oil/grease, and carry out the conductivity test, while the other sample was fixed by concentrated HNO_3 (nitric acid). Oil and grease in the water samples were analysed by extraction of TPH (total petroleum hydrocarbon). Selected heavy metals in the groundwater samples – arsenic, cadmium, cobalt, chromium, copper, manganese, nickel, lead, selenium, and zinc – were also studied.

At zero day, before the implementation of the project, ETP sludge samples were collected from 10 random points on two blocks of the bioremediation site for analysis. ETP sludge from Block A was found to

Figure 2
Application of oilzapper at the sludge dumping site

contain 38.65% solvent extractable TPH, 28.13% moisture/water content, 28.10% organic heavy fraction of hydrocarbon, and 5.12% sediment/inorganic waste. Block B contained 42.77% solvent extractable TPH, 24.05% moisture/water content, 26.47% organic heavy fraction of hydrocarbon, and 6.71% sediment/inorganic waste.

ETP sludge samples were regularly taken from both blocks on the site. As shown in Table 1, the first set of samples (on zero day) was taken on January 2006, while the final samples were collected in June 2006. Table 1 and Figure 3 show the TPH content and percentage of biodegradation over this period.

The results also showed that the concentration of heavy metals in the soil at the bioremediation site was also within permissible limits as per the Hazardous Wastes Management and Handling Rules (2003) as seen in Table 2.

TABLE 1
TPH CONTENT AND PERCENTAGE OF BIODEGRADATION OVER FIVE MONTHS (JANUARY–JUNE 2006)

Sample	*Sampling period*	*Sampling time*	*Block A TPH (%)*	*Biodegradation (%)*	*Block B TPH (%)*	*Biodegradation (%)*
1st sample	Zero day	January 2006	38.65	–	42.77	–
2nd sample	1 month	February 2006	29.76	23.00	33.45	21.79
3rd sample	2 months	March 2006	20.10	47.99	29.82	30.28
4th sample	3 months	April 2006	13.53	64.99	26.10	38.98
5th sample	5 months	June 2006	0.80	97.93	0.98	97.71

Zero day – before implementation of the project; TPH – total petroleum hydrocarbon content

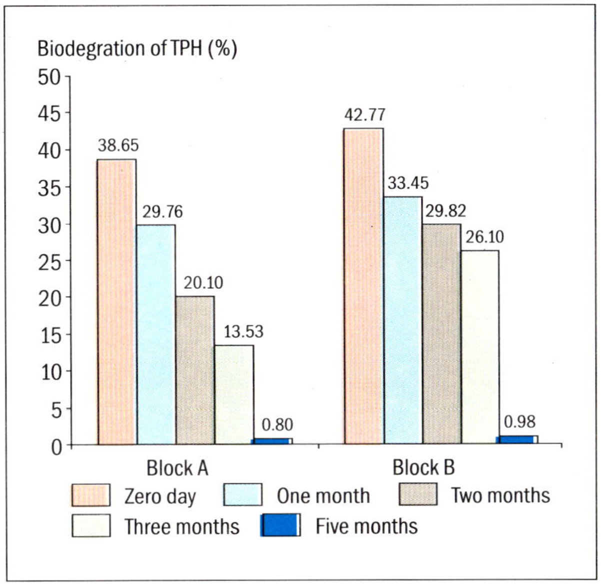

Figure 3
Total petroleum hydrocarbon content in Block A and Block B of the Lakwa bioremediation site over five months (January-June 2006)

TABLE 2
Selected heavy metal concentration in ETP oily sludge at the Lakwa bioremediation site

Heavy metals	*Concentration of heavy metals in bioremediation site*	*Permissible limit as per the Hazardous Wastes Management (2003) Amendment and Handling Rules*
Zn	1.2	20 000 PPM
Mn	27.4	
Cu	1.2	Total concentration
Ni	2.83	5000 PPM
Pb	BDL	
Co	–	
As	1.82	Total concentration
Cd	BDL	50 PPM
Cr (total)	0.85	50 PPM
Se	BDL	

BDL - below detection level by AAS (atomic absorption spectrophotometer);
PPM - parts per million; Zn - zinc; Mn - manganese; Cu - copper; Ni - nickel; Pb - lead; Co - cobalt; As - arsenic; Cd - cadmium; Cr - chromium; Se - selenium

Benefits from the project

The results proved that oilzapper can not only degrade all constituents of crude oil to over 90%, but it is also an ecologically sound and natural process. It is also faster as compared to natural degradation. Further, the treated soil, which is rich in biomass and organic carbon, can be used or sold as manure for greenbelt development, routine agriculture, etc.

The oilzapper treatment at the Lakwa site also led to large cost savings for ONGC. While ONGC used to spend over Rs 500 million in the storage of sludge in polymer-lined secure pits and utilize large areas of land, treatment with oilzapper costs the company only Rs 2500 per m^3 of oil contaminated soil/oily sludge.

CASE STUDY

at a glance

- The annual turnover for the ONGC (Oil Natural Gas Corporation Ltd) Group for 2005/06 was Rs 864.14 billion.
- Prior to the implementation of the project, ONGC was disposing off tank bottom oily sludge and oily effluent soaked soil waste in synthetic line pits–this approach utilized large areas of land, led to seepage, often caused overflows during the rainy seasons, required high construction costs, and did not eliminate the pollutants.
- ONGC thus approached TERI, New Delhi, for a new method using bioremediation for the degradation of oily sludge, waste water effluent sludge, and other oily waste, using oilzapper culture, containing live sludge-degrading bacteria.
- Once applied on a given site, the microbes in oilzapper degrade the toxic hydrocarbon compounds of the sludge, leaving non-hazardous end products or metabolites that are harmless to the environment.
- ONGC used oilzapper at its Lakwa, Assam Asset, ETP (effluent treatment plant) sludge pit for the bioremediation of 966 tonnes of oily sludge.
- Regular sampling showed that oilzapper not only degraded all constituents of crude oil to over 90%, but was also an ecologically sound and natural process.
- Soil treated with oilzapper was rich in biomass and organic carbon, and fit for use as manure for greenbelt development, routine agriculture, etc.
- Oilzapper treatment at the Lakwa site also led to large cost savings as while ONGC used to spend over Rs 500 million in the storage of sludge in polymer-lined secure pits and utilize large areas of land, treatment with oilzapper cost the company only Rs 2500 per m^3 of oil-contaminated soil/oily sludge.

CASE STUDY

7

Reliance Industries Ltd, Hazira

PET WASTE RECYCLING PROJECT

Background

The Hazira Manufacturing Unit of RIL (Reliance Industries Ltd), hereafter called 'RIL Hazira', is located in District Surat, Gujarat. The major products manufactured at RIL Hazira are ethylene, propylene, PDEB (para di ethyl benzene), butadiene, poly ethylene, poly propylene, PVC (poly vinyl chloride), VCM (vinyl chloride monomer), MEG (mono ethylene glycol), PTA (purified terephthalic acid), POY (partially oriented yarn), PET (poly ethylene terephthalate), PSF (polyester staple fibre), FF (fibrefill), and poly ethylene duct. RIL Hazira had a turnover of over Rs 175 billion during 2005/06.

Environmental profile

RIL Hazira's EMS (environment management system) is ISO 14001 certified, and the company has received several awards—those received in recent times include the Award for Excellence in Energy Conservation by the CII (Confederation of Indian Industries) in 2006, the Golden Peacock Eco-innovation Award by the World Environment Foundation in 2003, the National Energy Conservation Award by the Ministry of Power, Government of India, in 2002, and the Corporate Environmental Award by TERI in 2002.

RIL is committed to the abatement of water and air pollution and also has an effective hazardous waste management system in place. The PVC and PTA plants have fogging nozzles that have significantly brought down SPM (suspended particulate matter) levels. RSPM (respirable suspended particulate matter) dust concentration in ambient air is monitored and kept within the statutory limit. To keep SO_2 (sulphur dioxide) levels to the minimum, sulphur-free natural gas is used to the greatest extent possible. Low NO_x burners are installed in furnaces of the cracker plant and DTA vapourizers. To control fugitive emissions, better environmental measures such as provision of double mechanical seal, closed loop sampling, online computerized detection system, and benzene vapour recovery system are in place.

Measures taken to abate water pollution include automation of the ETP (effluent treatment plant), pure oxygen aeration in the aeration tank, effluent recycling scheme for recycling the treated effluent as cooling tower make-up, enhancement of sludge-drying capacity through introduction of hot-air-based sludge-drying system, replacement of the steam-based ejector system with a glycol-based ejector system that reduces effluents and COD (chemical oxygen demand) load, etc. The major categories of hazardous waste and their management practices are highlighted in Table 1.

RIL has an environmental risk management system and disaster management plan, and takes numerous steps towards resource conservation, especially energy conservation.

The company is dedicated to CSR (corporate social responsibility) and has a detailed CSR policy based on seven pillars: (1) education outreach

TABLE 1
Hazardous waste management

Solid waste	*Management practice*
ETP sludge	Sent to a GPCB-approved site for secured landfilling
Spent catalysts	Sold to a party with valid license from SPCB for recovery of metals
PTA incinerator ash	Sold to a party that recovers cobalt and manganese, prepares cobalt acetate and manganese acetate solution, and sells it back to the company
Coke	Used as fuel by a party with GPCB license
Discarded containers	Decontaminated containers are sold to a having valid license of GPCB
Spent batteries	Lead acid batteries are sold to a CPCB approved recycler. Nickel cadmium batteries are bought back by the supplier

GPCB – Gujarat Pollution Control Board; SPCB – state pollution control board; ETP – effluent treatment plant; PTA – purified terephthalic acid; CPCB – Central Pollution Control Board

programmes that provide training to teachers, organize distance learning programmes, give grants to schools, start schools in the vicinity, etc.; (2) safety outreach programmes such as the setting up of the Driver's Safety Training Centre in September 2005, and the road accident prevention mission; (3) quality of life outreach programmes that focus on infrastructure, health-care, rescue, and relief activities; (4) ecosystem outreach programme including its 'Vermicompost project', the 'Reliance Tapi river front development project', and the 'Diesel from jatropha' project; (5) resource efficiency outreach programme including rainwater harvesting systems, drip irrigation schemes, and projects for fuel cells and solar-system-based houses, etc.; (6) talent nurturing and self-sufficiency outreach programmes for school children; self-help groups have been established in the local community; a Special Olympics organized for the mentally challenged, physically disabled, and deaf and dumb, etc.; and (7) responsible care and product stewardship outreach programmes that include the 'Recycling of PET waste' project that has led to the upliftment of PET ragpickers, discussed in the case study.

Further, RIL Hazira has started project development under the CDM (clean development mechanism) as a part of its commitment to mitigate global warming. As part of the CDM movement, the company is spreading awareness among employees on the Kyoto Protocol and CDM, assessing all environmental projects for carrying out the potential CDM projects, and developing two projects worth 24 000 CERs (certified emission reductions).

Case study

RIL's Hazira unit has the world's biggest polyester recycling facility and procures polyester and PET bottle waste from over 40 producers. Recycling both pre-consumer and post-consumer PET waste, that would take about 100 years to biodegrade, would lessen the amount of waste going into landfills, thereby reducing the amount of toxic chemicals absorbed into the earth. Thus, the facility not only eliminates polyester waste that endangers the environment, but also adds value to the waste by converting it into children's toys, pillows, furnishings, etc.

Figure 1 shows the process of turning polyester waste into fibrefill products at the PFF (polyester fibrefill plant). The PFF manufacturing process is depicted in Figure 2. The process entails four steps: glycolysis, continuous polymerization, spinning, and then drawing. The raw material

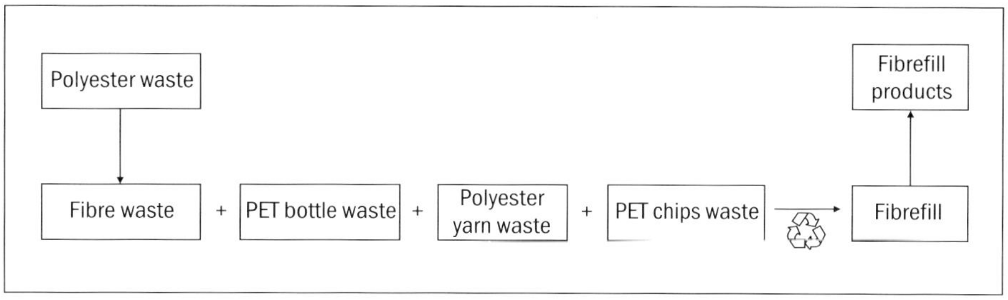

Figure 1
Polyester fibrefill plant process

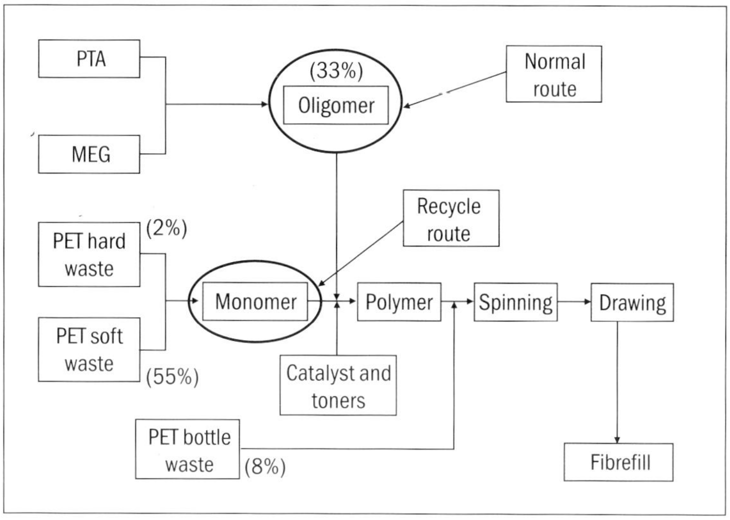

Figure 2
Polyester fibrefill manufacturing process

used in the process is polyester waste from RIL polyester plants and other polyester producers. In the glycolysis step, polyester waste is stocked, segregated, weighed, and charged into the glycolysis reactor for de-polymerizing it into a monomer by adding heat and catalyst. The monomer is then transferred to the continuous polymerization unit. At this stage, a stream of virgin oligomer joins the monomer stream to improve the product quality and plant yield. Catalysts, delusterant agents, toners, and optical brighteners are added. The combined streams of monomer, oligomer, and additives then go to the continuous polymerization section, where high-viscosity polymer is formed at a high temperature and under vacuum conditions. The polymer is pumped into the spinning section where it is filtered and extruded through the tiny holes of the spinneret. Filaments emerging from the spinneret are solidified by cold air and the filament bundle is finally received from the spinning and drawn, crimped, heat set, and then cut into fibre of the desired lengths as per market demand.

The project implemented by RIL Hazira aimed not only at maximizing the recycling of polyester waste, but also at increasing the overall production capacity of fibrefill and reducing its production cost by minimizing the consumption of oilgomer, toners, catalysts, and other high-cost items. The project was designed in four phases over four years, 2002/03–2006/07. The four phases of the project entailed the following.

1 Revamping of the PFF project
2 Commissioning of the post consumer use PET bottles recycle system
3 Increasing water recycle quantity with a corresponding decrease in virgin through a Six Sigma project
4 Maximizing waste recycle in all forms

Revamping the PFF plant resulted in an increase in production capacity from 66 MTD (metric tonnes per day) to 82 MTD. The new system entailed the following.

- Installation of 16 extra spinning positions from where polymerized polymer is extruded and spun into fibre
- Installation of new polymer cooler in the polymer line
- Modification in the monomer filtration system
- Solid waste usage in the project
- Use of larger crimper rolls for increased throughput in the drawline

Post consumer use PET bottles were seen as an alternative source that could be recycled in the plant and hence, a system was developed to recycle this bottle waste.

- A separate agency for the collection, procurement, sorting, crushing, cleaning, and washing of dirty bottle flakes was set up to make it fit for recycle use.
- A system was designed in-house for the drying, crystallization, extrusion, and filtration of the polymer generated from recycled bottle flakes.
- The flakes are passed through metal detectors and stored in a silo; these are then dried and crystallized with dry and hot air.
- The dried flakes are then melted, extruded, filtered, and injected into the main transfer line.

Since the commissioning of the system, approximately 2000 MT (metric tonnes) of used PET bottles have been recycled. Several modifications have been carried out to further improve the reliability of the system and further increase the PET waste bottle processing rate to about 6 TPD. Improvements in terms of production, plant yield, and quality were carried out under phase 3 of the project, using Six Sigma methodology.

The utilization of used PET bottles by RIL Hazira has also provided employment opportunities for numerous slum-dwellers. Sourcing PET bottles from ragpickers has increased their incomes by an estimated 150% and has eliminated middlemen and agencies from the process. Figure 3 shows the increase in PET bottle collection by ragpickers from September 2004 to November 2005; this trend is continuing to grow.

The commissioning of the post consumer use PET bottle recycling system required an investment of Rs 36.9 million and yearly operation

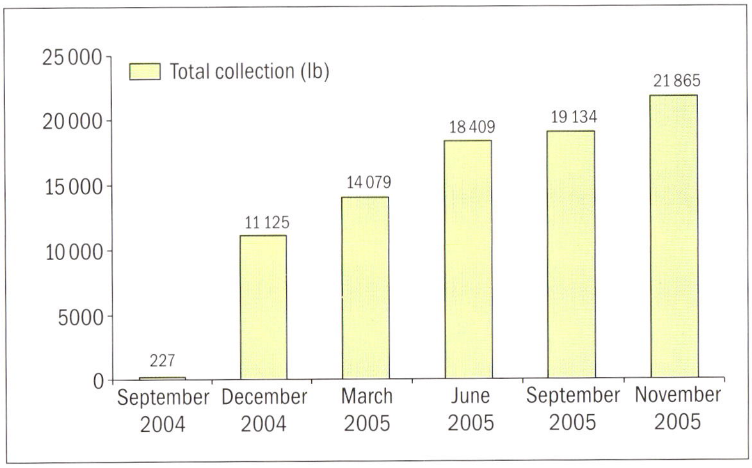

Figure 3
Increasing trend of PET bottle collection by ragpickers (September 2004–November 2005)

cost of Rs 600 000. However, the cost benefit from the project is Rs 55.2 million. Thus, the payback period for investment in the project is only eight months. Phase 1 of the project has brought in a benefit of Rs 52 million and a 21 month payback period, while phase 3 generates monetary benefits of Rs 70.3 million against yearly operation cost of Rs 2 million and an investment of Rs 500 000. Phase 4 is expected to bring benefits of Rs 20 million against yearly operation cost of Rs 500 000.

Non-monetary benefits include value-addition to a potentially dangerous pollutant, environment protection, additional employment generation, and improvement in the living standards of ragpickers.

RIL Hazira, in the fourth phase of the project, aims to further improve the system by maximizing the recycling of all forms of waste. Plans are in place to further decrease the 25% virgin oligomer component while increasing production volumes.

CASE STUDY

at a glance

- RIL (Reliance Industries Ltd) Hazira Manufacturing Unit, located in Surat, Gujarat, had a turnover of over Rs 175 billion during 2005/06.
- RIL Hazira has the world's biggest polyester recycling facility and procures polyester and PET (poly ethylene terephthalate) bottle waste – that can take about 100 years to biodegrade – from over 40 producers.
- The unit implemented a project aimed at maximizing the recycling of polyester waste, increasing the overall production capacity of fibrefill, and reducing its production cost by minimizing the consumption of oilgomer, toners, catalysts, and other high-cost items.
- A four-phase project was designed for the period 2002/03–2006/07.
- The first phase of the project entailed the revamping of the PFF (polyester fibrefill) project and resulted in an increase in production capacity from 66 MTD (metric tonnes per day) to 82 MTD.
- The second phase was the commissioning of post consumer use PET bottles recycle system, through which approximately 2000 MT (metric tonnes) of used PET bottles have been recycled and ragpickers have provided employment opportunities.
- The third phase entailed the increase in water recycle quantity with a corresponding decrease in virgin through a Six Sigma project.
- The fourth and final phase was used for the maximization of waste recycle in all forms.
- Benefits from the project include value-addition to potential pollutants, environment protection, additional employment generation, and improvement in the living standards of ragpickers, in addition to sizeable monetary savings for RIL.

CASE STUDY

8

Reliance Industries Ltd, Patalganga Manufacturing Complex

REDUCTION IN THE VOLUME OF HAZARDOUS WASTE USING WASTE HEAT, RESULTING IN ENVIRONMENT-FRIENDLY HANDLING

Background

RIL (Reliance Industries Ltd), the flagship company of the Reliance Group, is a Fortune Global 500 company and India's largest private sector company. It was the first Indian private sector company to exceed $2 billion in net profit. RIL's net profit for 2005/06 was Rs 90 billion, and the annual turnover recorded for the same period was over Rs 891 billion.

Major products manufactured at RIL's Patalganga Manufacturing Complex, located in the Raigad district of Maharashtra, include PX (paraxylene), PTA (purified terephthalic acid), POY (partially oriented yarn), FDY (fully drawn yarn), SDY (semi-drawn yarn), POY chips, PSF (polyester staple fibre), LAB (linear alkyl benezene), and NP (normal paraffin).

Environmental profile

RIL-Patalganga has an environment policy and an environment cell at both the corporate and facility levels. It complies with all the requirements of ISO 9000, ISO 14000, and OHSAS 18000. All the manufacturing processes at the site have been designed to achieve the highest standards of process safety, raw material and resource consumption, and eco-efficiency.

Effective measures and high-end technology are used at the Patalganga facility to minimize or abate pollutants. SPM (suspended particulate matter) emission in the PTA plant is reduced through the installation of the new CTA (crude terephthalic acid) dryer with a central discharge system to reduce powder leakages and through provision of a hood at the

Figure 1
Reliance Industries Ltd, Patalganga

purification dryer discharge end. Dryer seal modification and vent scrubber modification also help to control SPM emission. Bag filters to prevent dust pollution are regularly inspected for tears. To control SO_2 (sulphur dioxide), low-sulphur fuel is used, process off gas and biogas generated in anaerobic digesters of the PTA plant effluent treatment facilities is utilized, and VOCs (volatile organic compounds) generated in the polyester process organic strippers are fired as fuel in the dow vapourizer for the dow vapourizer polyester plant. Low NO_x (oxides of nitrogen) burners are installed in the heaters of the PX plant.

To treat process waste water, RIL-Patalganga has two ETPs (effluent treatment plants) for the polyester division and PTA division (Figure 2). Disposal practices are adopted for hazardous wastes generated on site. All ETP sludge generated is disposed off at MWML's (Mumbai Waste Management Ltd) landfill site; used/waste oil is sold to various used/ waste oil processors; and incinerator ash is sold to various metal recovery units for the recovery of Co (cobalt) and Mn (manganese).

Energy consumption is regularly monitored and has, as shown in Figure 3, been consistently decreasing since 2002 due to energy conservation measures. Energy conservation by heat integration and superior equipment design is achieved at the design stage of all plants.

RIL-Patalganga has several precautionary measures in place in case of hazards in handling chemicals. The bulk of raw materials is transported through underground pipelines rather than road routes. There is an on-site disaster management plan, fire protection system, MARG (mutual aid response group), spill prevention and response plan, and round-the-clock medical facilities on site. A high-velocity water spray system for HF (hydrofluoric) acid handling area, linear heat detection system in naphtha and PX tanks, auto-deluge system for LPG (liquefied petroleum gas) loading gantry and pumps, and PPM and LEL (lower explosive limit) detectors at strategic locations to detect even the smallest leakage of toxic

Figure 2
Effluent treatment plant

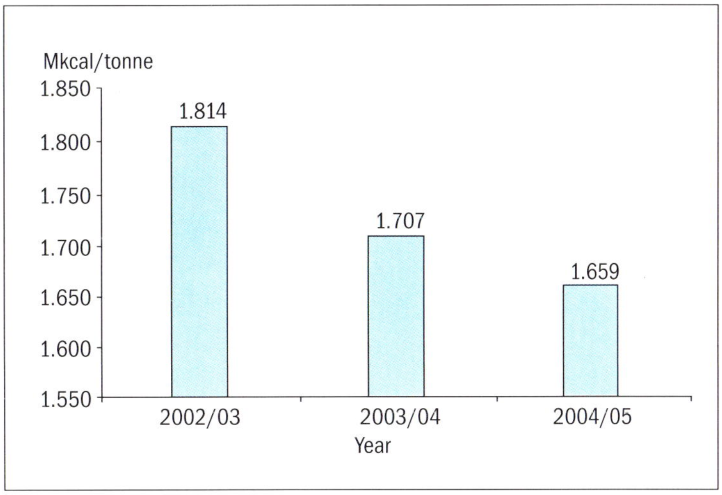

Figure 3
Energy consumption (2002-05)

and combustible gases are some precautionary and safety technologies utilized. Protective appliances for personnel include on-line breathing apparatus sets, self-contained breathing air sets, escape sets, dust masks, dust respirators, helmets, PVC (polyvinyl chloride) goggles, panoramic goggles, safety shoes, ear-plugs and ear-muffs, hand gloves, asbestos suits, fire proxy suits, etc.

The facility has initiated a CDM (clean development mechanism) project titled 'Demand-side energy efficiency at RIL-Patalganga' that uses flared biogas in the process heater to reduce fuel oil consumption, and has installed a variable speed screw compressor for compressed air to reduce power consumption and hence fuel in the CPP (captive power plant).

RIL has an extensive social responsibility and community development programme. HIV/AIDS education drives are regularly conducted for drivers, cleaners, and other staff at the facility, and blood donation camps are organized regularly. Several activities and training programmes are undertaken at the facility to spread awareness on safety issues for employees. Training programmes are also held at RIL-Patalganga for college students.

Case study

Spin flash dryer at energy centre

The ETP at RIL-Patalganga's energy centre generates about 100 MT (metric tonnes) of bio-sludge with 16%–20% solids through its BFP (belt filter press) every month. The bio-sludge is usually stored on sludge-drying beds and disposed off to MWML at the rate of Rs 2225/MT. Another alternative method to dispose the sludge is to incinerate it with furnace oil—this procedure is not only costly for the company, but also has adverse impacts on the environment.

In January 2005, after exploring the possibility of heating bio-sludge and drying it up to 96% solids by utilizing hot waste gases of the HRSG (heat recovery system generator) stack, a sludge-drying unit was installed on site, as shown in Figure 4. Hot gases in the stack dry the bio-sludge up to 96% solids with a net reduction of volume to one tonne per day. The dried material can then be processed for metal recovery and subsequently be safely disposed off.

Figure 4
Flash dryer at the energy centre

Process description

Sludge is transported from the ETP dryer area and unloaded in a bin. Twin screws are provided at the bottom of the bin to partially condition and discharge the sludge on a conveyor belt, which, in turn, will discharge the sludge into the feed bin of the flash dryer. Twin screws are provided at the bottom of these bins to feed the material in the agitated flash dryer. The speed of the feed bin screw is regulated to maintain outlet temperature at set point.

Hot air is taken from the HRSG stack after it passes through the economizer. Hot gases enter the air distributor, and the wet sludge in agitation comes in contact with the hot air and dries in the bottom section. The dried sludge powder is carried with drying air and separated in the reverse pulse jet bag filter. It is discharged through a valve at the bottom of the product bin. Clear exhaust air is then vented by a blower through the exhaust duct to the HRSG stack. The entire operation is controlled through a PLC (programmable logic controller) in the local operating panel.

Benefits

The principle advantages of the proposed evaporation of sludge using heat available with flue gases of the flash drier result in, first, lower volumes of sludge for disposal. It is expected that the disposal of ETP sludge will reduce from 1000 MT/annum to 150 MT/annum. Second, it lowers the footprint of hazardous wastes from the PTA manufacturing plant. Third, it allows efficient energy utilization of flue gases from another section of the plant, resulting in lower thermal pollution from the CPP. The temperature of flue gas from the waste heat recovery system has reduced from 210 °C to 95 °C. Moreover, as the company no longer has to dispose off the ETP sludge to an external landfill site, the new method results in an annual cost saving of Rs 52.1 million, as against the cost of the system of Rs 14.6 million. The period of payback on investment, therefore, is only five months.

CASE STUDY

at a glance

- RIL (Reliance Industries Ltd) is India's largest private sector company, with an annual turnover of Rs 891 billion (for 2005/06).
- The ETP (effluent treatment plant) at RIL's Patalganga Manufacturing Complex, located in the Raigad district of Maharashtra, generates about 100 MT (metric tonnes) of bio-sludge with 16%–20% solids through the BFP (belt filter press) every month.
- A flash drier was installed on the site to heat the bio-sludge and dry it up to 96% solids by utilizing hot waste gases of the HRSG (heat recovery system generator) stack; the dried material can then be processed for metal recovery and subsequently be safely disposed off.
- The flash drier is expected to reduce the ETP sludge from 1000 MT/annum to 150 MT/annum.
- The temperature of flue gas from the waste heat recovery system has reduced from 210 °C to 95 °C, resulting in lower thermal pollution and hence leading to energy efficiency.
- Finally, the company saves up to Rs 52.1 million each year, as it no longer has to dispose off ETP sludge at the external landfill sites.

CASE STUDY

9

Tata Sponge Iron Ltd

CHANGING OF WET GAS CLEANING PLANT TO ELECTROSTATIC PRECIPITATOR AT KILN-I AND GENERATION OF POWER BY UTILIZATION OF HEAT FROM WASTE GAS

Background

TSIL (Tata Sponge Iron Ltd) is located at Joda, Orissa, and produces 390 000 MT (million tonnes) of sponge iron and 26 MW (megawatts) of power per annum. The company has an annual turnover of over Rs 2 billion. The TSIL township is depicted in Figure 1.

Environmental profile

TSIL has an environmental policy at the company level and is the first sponge iron plant in the country to have received ISO 14001 certification. It has received several environmental awards in the past including the Best Performance in Safety and Environment Management Award by the Directorate of Factories and Boilers, Government of Orissa in 2005, the first prize for the TERI Corporate Environment Awards in 2002/03, and a Platinum-rating for the Greentech Environment Excellence Award by the Greentech Foundation in the same year. Other awards include the first prize in the Quality Competition by the CII (Confederation of Industries) in 2003, the National Safety Award by the Government of India in 2003, and the State Safety Award by the Government of Orissa in 2003.

While sponge iron industries are known to be highly polluting, TSIL aims to stand out as an exception by being a leader in the field of the environment. The company has introduced various facilities and equipment to minimize or abate pollutants. Operations like crushing, screening, and conveying raw materials to produce sponge iron result in the generation of large amounts of dust which pollutes the environment. Hence, separate ESPs (electrostatic precipitators) are provided at each of the kilns along with a DE (dust extraction) system. A dust suppression system through chemical spraying is in place in raw material circuits. Moreover, a telescopic unloading spout is installed at the char and sponge

Figure 1
Tata Sponge Township in 2005

iron unloading areas to avoid the flying of fugitive dust. For the suppression of dust, regular truck tyre cleaning is carried out and water is sprayed onto roads and concrete floors. Further, a vacuum machine is used for road cleaning. These are used on a regular basis at the TSIL campus, and parks and gardens are well maintained. There are on-site disposal yards for degradable and biodegradable wastes. These measures have helped keep SPM (suspended particulate matter) at about 54% of the SPCB (state pollution control board) norm of 500 $\mu g/m^3$(micrograms per cubic metre). Dust and noise mapping are also done to eliminate/ minimize dust and noise. Other air pollutants like SO_2 (sulphur dioxide) and NO_x (oxides of nitrogen), are produced in small concentrations and fall within the levels set by the SPCB—20 $\mu g/m^3$ and 40 $\mu g/m^3$, respectively.

The company does not discharge any waste water outside its premises. All the water is recirculated in the process. Regeneration water of the power plant is recycled after the settling of TSS (total suspended solids) and maintenance of pH. Water used in kitchens and bathrooms is charged to the ground by passing it through soak pits.

TSIL generates about 20 kL (kilolitres) of used oil per annum, which is sold to parties authorized by the CPCB (Central Pollution Control Board) or MoEF (Ministry of Environment and Forests). The company takes numerous measures to monitor equipment and EMPs to keep the consumption of oil and lubricants to their minimum, in order to generate a minimal amount of waste. In 2005/06, consumption in lubricant oil LC-1 was reduced by 49% and EP-2 was reduced by 71% due to the implementation of various EMPs.

Coal and iron fines are other industrial solid wastes generated during the production process—the full amounts of both are sold to private entrepreneurs or traders. Back spillage is reused inside the kiln after screening, and rejected spillage is sold. ESP (electrostatic precipitator) dust is used for the manufacture of bricks that are utilized for construction in the company. Coarse char and fine char are currently dumped in the waste yard; however, TSIL is exploring possibilities of utilizing the same along with coal fines in the fluidized bed boiler for the generation of power. Dedusting sludge, waste sludge coal shale, stone, etc. are partially sold to outside parties, and the remaining quantity is used for land reclamation.

Further, resource conservation initiatives include power generation from waste heat, recirculation of waste water, and full utilization of ESP dust to manufacture fly ash bricks.

TSIL's waste heat recovery project is a CDM (clean development mechanism) project, through which it aims to reduce 317 624 tonnes of CO_2 emission over a 10-year crediting period. The project is further discussed in the case study.

Case study

Replacement of wet gas cleaning plant to electrostatic precipitator

The company's wet GCP (gas cleaning plant) was installed in 1986, wherein hot gas was getting quenched and cleaned by water. A large quantity of water was being consumed in the cleaning of the gas and GCP sludge (waste) was settling in settling tanks, which was then transported to the dump yard. These activities were incurring a lot of expenditure and manpower, and were non-value additions. Moreover, the cleaning of the sludge ponds by men used to be painstaking and unhygienic.

Despite the installation of machines to handle the entire process, and the elimination of manpower from the operation, the problem of pollutants being generated persisted. Thus, during 2005/06, the company decided to replace its wet GCP in kiln-1 (Figure 2) with a WHRB (waste heat recovery boiler) followed by an ESP (Figure 3), which not only generates 7.5 MW (megawatts) of power, but also eliminates the requirement of a huge quantity of water for gas cleaning and the

Figure 2
Wet gas cleaning plant previously used

Figure 3
Electrostatic precipitator in use presently

generation of GCP sludge. Further, waste dust produced by the ESP could be fully utilized in the manufacture of bricks—almost 20 000 MT of fly ash is utilized each year (Figures 4a and 4b). Stack emission from the ESP is also less as compared to the wet gas cleaning system.

The environmental benefits achieved by the transformation of the wet GCP to WHRB followed by the ESP are shown in Table 1. Other savings include the following.

- Production savings as regular power failures used to hamper production
- Improvement in product quality due to regular power supply
- Water savings in the wet GCP process
- Heat savings from utilization of waste heat from the flue gas (discussed in next section)

Figures 4a and 4b
Brick manufacture from fly ash

TABLE 1
Environmental benefits from technology change

Parameters	*Kiln-1 with GCP (before)*	*Kiln-1 with WHRB and ESP (after)*
Air emission level from stack	100–110 mg/Nm^3	40–50 mg/Nm^3
Volume of gas emission	154 000 Nm^3/h	130 000 Nm^3/h
Generation of waste water	500 kL/day	Nil
Generation of solid waste	50 MT/day of waste sludge	40 MT/day of dust useful for brick-making
Generation of power	Nil	7.5 MW

GCP – gas cleaning plant; WHRB – waste heat recovery boiler; ESP – electrostatic precipitator; kL – kilolitres; MT – million tonnes; MW – megawatt; Nm^3 – normalized cubic metre

- Elimination of GCP sludge
- Land previously used for dumping waste sludge

Generation of power by utilization of heat from waste gas

The lack of stable power supply at TSIL and increasing power tariffs were directly affecting the productivity and operating costs of the unit. In order to resolve these problems, TSIL installed three DG (diesel generator) sets, each with a capacity of 1000 kV (kilovolts), to run the kilns.

This solution, however, did not fully address the total operational requirements, and TSIL carried out extensive brainstorming sessions, and engineering and commercial project evaluations. After much research and evaluation, TSIL decided to install a waste heat recovery boiler (Figure 5) to generate power from waste heat of the rotary kiln, wherein no other natural resources were required, except water for running the DM plant. This not only utilized the sensible heat of waste gas for the generation of steam, but also utilized the heat released due to the burning of unburnt carbon available in the hot exhaust gas to maintain its temperature as the unburnt carbon gets much time and space for burning. This also reduces the carbon load of the exhaust gas to the ESP.

The initial power plant installed by the company in December 2001 had a capacity of 7.5 MW and utilized waste heat of the flue gas of kiln-2. This generation of power from waste gas not only solved the problem of power interruption, but also reduced operation costs, as the generation was meeting the total requirement of the plant, and was generating surplus power of about 2 MW that was being sold to GRIDCO (Grid Corporation of Orissa Ltd).

The generation of power through the 7.5-MW power plant of kiln-2 has, from 2003 to 2006, resulted in savings in coal consumption of about 125 450 MT, which translates into a cost saving of approximately Rs 250 million to the company (as seen in Table 2). Reduction of coal consumption also reduces the release of CO_2 into the atmosphere.

In 2005/06, the company set up a third kiln (kiln-3) that had a capacity of 500 TPD (tonnes per day) with a power plant capable of generating 11 MW power from waste gas. As a result, the total generation of power

Figure 5
Waste heat recovery boiler

TABLE 2
Coal savings through the installation of 7.5-MW power plant

Year	*Total power generation* (kWh)	*Coal saved* (kWh x 0.9/1000)	*Saving value* (at Rs 2000/MT)
2003/04	38 177 000	34 359 MT	Rs 68.7 million
2004/05	44 986 600	40 488 MT	Rs 80.9 million
2005/06	40 210 700	36 190 MT	Rs 72.3 million

kWh – kilowatt-hour; MT – million tonnes

by the company increased to 26 MW—as only 9 MW of total generation was required by the company, the remaining 17 MW was sold to a third party. Further, the emission level of the ESP discharge stack was reduced by over 50%.

CASE STUDY at a glance

- TSIL (Tata Sponge Iron Ltd), located at Joda, Orissa, produces sponge iron and power, and has an annual turnover of over Rs 2 billion.
- In order to resolve the problems of large quantities of waste water generation, interrupted power supply that was affecting production levels and quality, heat loss, and generation of GCP (gas cleaning plant) sludge waste, the company introduced innovative technologies.
- The wet gas cleaning plant was replaced by a WHRB (waste heat recovery boiler) followed by an ESP (electrostatic precipitator), which not only generates 7.5 MW (megawatts) of power, but also eliminates the requirement of a huge quantity of water and eliminates GCP sludge.
- A waste heat recovery boiler was also installed to generate power from the waste heat of the rotary kiln–this was also effective in utilizing unburned carbon from the kiln.
- The case study proves not only to be financially viable, but has also resulted in decreased pollution of the environment.
- The transformation of the wet GCP to the dry ESP in kiln-1 has eliminated the water requirement of cleaning and quenching the waste gas; and the heat of the gas is used to generate steam for the power plant.
- The installation of the ESP has led to a reduction in emission level particulate matter from over 100 mg/Nm^3 (using the wet GCP) to under 50 mg/Nm^3.
- Wet GCP sludge (waste) has been completely eliminated, and the entire quantity of ESP dust collected is used in the manufacture of bricks.
- The utilization of waste heat to generate power saves almost 150 000 MT of coal, a non-renewable natural resource, and hence, significantly lowers the release of CO_2 to the atmosphere.
- Total power generation by the company increased to 26 MW–17 MW more than the company's own requirement.

CASE STUDY

10

UAL–Bengal, UAL Industries Ltd

ZERO WASTE, ZERO EFFLUENT, AND ECO-FRIENDLY MANAGEMENT SYSTEM IMPLEMENTATION AT THE FIBRE CEMENT SHEET INDUSTRY

Background

UAL (formerly, Utkal Asbestos Ltd)–Bengal is one of the leading manufacturers of fibre cement corrugated sheets, building boards, moulded goods, and roofing accessories. The company sells its products under the brand name 'KONARK'. Its manufacturing plant, situated 25 km from Kharagpur, in the West Midnapore district of West Bengal, achieved an annual production level of 140 000 MT (million tonnes), as on 31 March 2005. UAL–Bengal realized sales of Rs 966.3 million for the same period.

Environmental profile

UAL–Bengal has an environmental policy and an environmental department existing at both corporate and facility levels. The environment policy is included under the company's written Integrated Management System Policy, along with its quality and occupational health and safety policies. UAL–Bengal has been awarded ISO 9001: 2000 for its QMS (quality management system), ISO 14001: 2004 for its EMS (environmental management system), and OHSAS 18001: 1999 for its occupational health and safety management system.

The organization conducts several programmes on environmental awareness for employees and the general public. In addition to organizing employee training programmes, it also celebrates World Environment Day, Forest Week, and Safety Week every year.

Social welfare activities undertaken by the company include the construction of a primary school in a village; distribution of clothes and blankets to leprosy patients at a facility near the factory; provision of fibre sheets for roofing of shelters and schools for orphans and handicapped children; provision of jobs for the poor, underprivileged, and widowed women in the factory; provision of area with medical and ambulance facilities, etc. A rooftop rainwater harvesting project is also under way for recharging water into the subsoil to increase the groundwater level. UAL–Bengal has planted over 25 species of plants in a number of schools, over 40 varieties on the Guptamani–Kultikri Road, and over 80 on its premises. The company also uses fly ash, a waste material of thermal power plants, in its process of producing asbestos cement sheets.

The manufacturing facility emits the following air pollutants: CO (carbon monoxide), CO_2 (carbon dioxide), asbestos fibre concentration per cm^3 of air, and dust concentration. These are minimized through regular checking and maintenance of pollution devices, ID (induction draft) fans, and PCD (pollution control device) components. Energy meters have also been installed with various PCDs. Dust collectors, an industrial vacuum cleaner, industrial floor swipper, and wet-mopping are also used to minimize the pollution load. Dust collected in the hoppers is reused in the process. The height of stacks is kept within the WBPCB (West Bengal Pollution Control Board) guidelines.

The company operates on the principle of 'zero effluent management', and hence, does not discharge any process effluents. A bore well is the source of water, and periodic testing is done for various parameters of water. The major categories of hazardous waste, all of which are totally recycled back into the process, are green waste, waste from bag-filter, and waste from salvaging, trimming, off-cuts and discards, and maintenance cleaning waste. Similarly, management practices and initiatives are in place for the reduction/recycling of other industrial wastes.

Case study

UAL–Bengal, in order to implement its 'zero waste, zero effluent, and eco-friendly management system' at its fibre cement sheet industry, has initiated a number of proactive measures such as introducing changes to its manufacturing process, managing its green belt, managing food waste, water conservation, and energy conservation.

The manufacturing process

In order to minimize waste and abate pollution resulting from the manufacturing process, UAL–Bengal has installed a number of PCDs (such as microprocessor-based pulse jet bag filters and wet scrubbers) at appropriate locations, and have connected them to suction hoods installed at various dust generation points. Any dust generated is thus sucked into the suction hoods and collected in bag filters of the PCDs; only clean air is released into the atmosphere. The PCDs are connected with energy meters to monitor and ensure that they are always in running condition; all PCDs are interlocked with the manufacturing process such that the process automatically stops if any of the devices is not functioning. Any waste generated from the PCDs is automatically collected in HDPE (high density polyethylene) bags and recycled back into the process.

Further, a fully-enclosed, fully-automatic PLC (programmable logic controlled) fibre-handling system with an automatic bag-opening device and a fibre-bag shredding machine is installed, ensuring that the filled-up fibre bags are cut in an enclosed chamber and the empty fibre bags are shredded in the shredding machine and recycled back into the process. This process requires no manual operation.

Adhering to the 'zero effluent management' concept, UAL–Bengal does not discharge any process effluent outside the plant premises. Waste water generated during production is continuously recycled back into the manufacturing process. Domestic waste water is discharged in a 30-kL (kilolitre) storage tank and reused in the development of the green belt, after preliminary treatment.

Under its 'zero waste management' principle, the company discards fibre cement sheets that would otherwise be termed as solid waste, by pulverizing and recycling them back into the manufacturing process. A pulverizer machine with an integrated bag filter and wet scrubber is installed. All categories of hazardous waste generated are totally recycled back into the process as described below. Green waste and waste from bag filters are fully recycled into the process, while waste from salvaging, trimming, off-cuts and discards, and maintenance cleaning waste is pulverized and recycled back into the process. About 1 MT/annum of machine scrap waste, which cannot be used, is sealed in HDPE bags and kept in a WBPCB (West Bengal Pollution Control Board)—authorized waste disposal pit. The action plan for the collection and disposal of non-hazardous solid waste is depicted in Figure 1, which summarizes the various waste generation and disposal

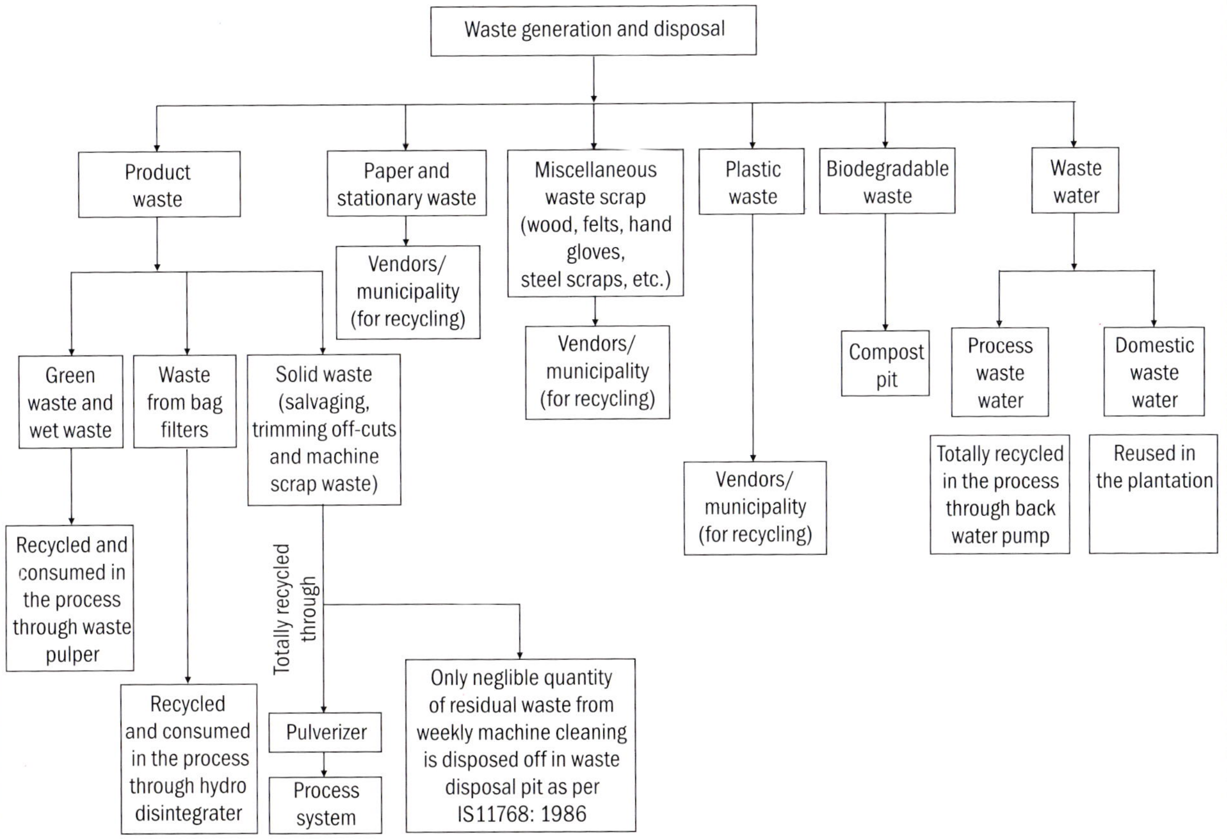

Figure 1
Non-hazardous waste generation and disposal

techniques at the company. Figure 2 shows the reduction of solid waste generation (by year), against the increasing production levels of the company.

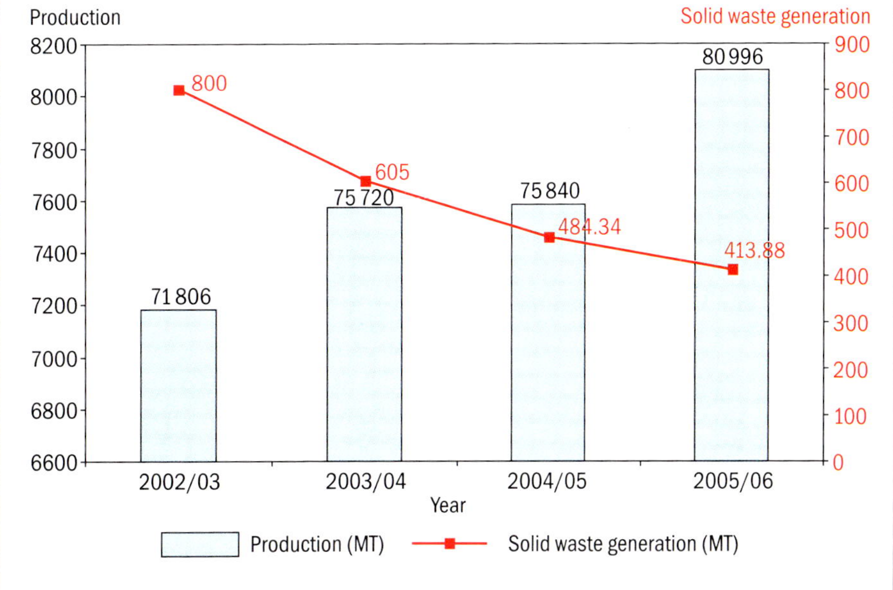

Figure 2
Production versus solid waste generation

Finally, UAL-Bengal upgrades its machines and equipment on a timely basis and is constantly trying to improve its practices and procedures.

Green belt management

UAL–Bengal has a green belt development plan through which it maintains and improves the ecological balance of the area. A total of 11 066 healthy plants are grown over 4.91 hectares of project area. Further plantation is under way. UAL–Bengal has presently planted at least 80 varieties of saplings and maintains 8000 healthy trees in and around the factory. About 3000 tea plants and a variety of medicinal plants have been planted in the township area with the assistance of IIT (Indian Institute of Technology)–Kharagpur.

Food waste management

A compost pit has been constructed (Figure 3) inside the township premises with the assistance of IIT–Kharagpur. Food and kitchen waste is kept in a pre-composting pit for seven days to release the toxic fumes. Thereafter, according to the requirement, food waste is mixed with cow-dung, grass, and water hyacinth, for better composting and to maintain the 50% optimum moisture. It is then sent to the compost pit containing earthworms and cow-dung, after which it is reused as fertilizer in the plantations.

Water conservation

In order to conserve water, UAL–Bengal has installed water meters at suitable points to monitor water consumption and take remedial action, if necessary. Domestic waste water generated from the factory and rain water collected in rainwater harvesting tanks is reused in plantation. The company has completed a roof-top rainwater harvesting project with the assistance of IIT–Kharagpur. Rain water is bacteriologically pure, free from organic matter, and soft in nature, and thus, harvested rain water is used for drinking purposes or cultivation work.

Energy conservation

The company has initiated various measures for energy conservation. For regular monitoring and analysis, 68 energy meters have been installed. Regular energy audits are conducted by the National Productivity Council.

Figure 3
Compost pit for biodegradation of food waste through vermicomposting

Moreover, proper insulation has been provided in the hot chamber, where sheets are cured. Burners are automatically controlled by the installation of a sequence controller so that they can be stopped when the requisite temperature has been reached. Through the effective synchronization of the four DG (diesel generating) sets in the powerhouse, UAL–Bengal has been able to conserve fuel. DG operation was also reduced through the electricity supply line from the WBSEB, saving large quantities of diesel oil.

The implementation of wind turbine fans on the roof of the shop floor also led to energy savings. Previously, there were 20 roof extractor fans of 1.5 kW (kilowatt) capacity in the shop floor that were running continuously throughout the day, thereby drawing substantial amounts of energy. After the implementation of new wind turbine fans, UAL–Bengal has been able to save 21 600 kWh (kilowatt-hours) energy per month ($20 \times 1.5 \times 24 \times 30 = 21\ 600$ kWh). The unit cost of electricity being approximately Rs 4.5, the use of wind turbine fans has led to a cost saving of Rs 97 200 per month (21 600 kWh × Rs 4.5 = Rs 97 200).

CASE STUDY

at a glance

- UAL–Bengal is one of the leading manufacturers of fibre cement corrugated sheets, building boards, moulded goods, and roofing accessories, achieving sales of Rs 966.3 million, as on 31 March 2005.
- UAL-Bengal, in order to implement its 'zero waste, zero effluent, and eco-friendly management system' at its fibre cement sheet industry, has initiated a number of proactive measures.
- A number of PCDs (pollution control devices) and a fully-automatic PLC (programmable logic controlled) fibre-handling system with an automatic bag-opening device and a fibre bag shredding machine were installed to minimize pollution.
- Waste water generated during production is continuously recycled back into the manufacturing process, while domestic waste water is reused in the development of the green belt.
- All categories of hazardous and non-hazardous waste are either totally recycled back into the process or safely disposed off.
- Under its green belt management plan, UAL–Bengal has grown over 11 066 plants over 4.91 hectares of project area.
- A compost pit has been constructed for food waste management.
- Rainwater harvesting, installation of water meters, and reuse of waste water for plantation are some water conservation measures implemented.
- The setting up of energy meters, provision of proper insulation in the hot chamber, installation of sequence controller, effective synchronization of DG (diesel generating) sets, and the implementation of wind turbine fans in place of the roof extractor fans on the roofs of the shop floor are some steps taken towards energy conservation.

PROCESS IMPROVEMENT

11
Chennai Petroleum Corporation Ltd

12
Indian Farmers Fertiliser Cooperative Ltd, Aonla Unit

13
Indian Oil Corporation

14
Indo Gulf Fertilisers Ltd

15
Numaligarh Refinery Ltd

16
Oil and Natural Gas Corporation Ltd, Cauvery Asset (flare gas reduction)

17
Oil and Natural Gas Corporation Ltd, Cauvery Asset
(zero cement dust emission module)

18
Rashtriya Ispat Nigam Ltd

19
Sona Koyo Steerings Systems Ltd

20
Toyota Kirloskar Motors Pvt. Ltd

21
Vishay Components India Pvt. Ltd

CASE STUDY 11

Chennai Petroleum Corporation Ltd

Implementation of thermal DeNO$_x$ technology in major stacks of refinery-III

Background

CPCL (Chennai Petroleum Corporation Ltd) is a group company of the Indian Oil Corporation Ltd, formerly knows as MRL (Madras Refineries Ltd). CPCL has two refineries – one in Manali, Chennai, and the other in the Cauvery basin at Nagapattinam – and a combined refining capacity of 10.5 MTPA (million tonnes per annum). The company's major products include LPG (liquefied petroleum gas), motor spirit, superior kerosene, ATF (aviation turbine fuel), high-speed diesel, naphtha, bitumen, lube base stocks, paraffin wax, fuel oil, hexane, and petrochemical feed stocks. The company achieved sales of over Rs 211 billion for the year ending 31 March 2006.

Environmental profile

CPCL is ISO 9001: 2000, ISO 14001, and OHSAS 18001 certified. It has received several awards in 2005/06 for environmental management such as the Golden Peacock Special Commendation Award of the World Environment Foundation, the Greentech Social Responsibility Award, and the Ecology Leadership Award by the Rotary Club, Chennai.

CPCL has set up a number of facilities for abating air pollutants—namely, SO_2, (sulphur dioxide) NO_x, (oxides of nitrogen) SPM (suspended particulate matter), HC (hydrocarbons), and CO (carbon monoxide). There is a flare gas recovery system to recover HC from gases going to the flare system, sulphur recovery plant to recover elemental sulphur from acid gases and reduce SO_2 emissions, CO boiler to produce steam by combustion of FCC (fluid catalytic cracking) regenerator off-gases, floating roof, and secondary seals in crude and product storage tanks to control HC fugitive emissions. Moreover, stack heights were increased in 1997 to meet the minimum stack height requirements for better dispersion into the atmosphere. DeNO$_x$ technology was also developed to reduce NO_x emission from stacks of refinery-III, explained in detail in the case study.

CPCL generated over 5 million kL (kilolitres) of liquid effluents in 2005/06, of which 100% was either reused or recycled. Over 3 million kL was reused for development of the green belt and other peripheral applications, while the remaining 2 million kL was reused for process applications.

Hazardous waste generated in the refinery is predominantly spent, catalysts and oily sludge. CPCL undertakes sludge treatment, recovery of oil, and safe disposal of all accumulated sludge through mechanical treatment and bioremediation (initiated for 500 kL of sludge on an experimental basis in collaboration with TERI, New Delhi).

CPCL is involved with other CSR (corporate social responsibility) activities such as developing green belt over about 30 acres of land, launching of an environmental literacy drive in neighbouring villages, creating awareness among employees on water conservation through exhibitions, competitions, book distribution, etc. It also has on-site and off-site emergency preparedness plans.

The company has identified 20 potential CDM (clean development mechanism) projects, of which, three – recovery of hydrogen from off-gases of the refinery, use of merox treater for producing ATF instead of conventional hydro-treating, and setting up of windmills – are in the initial stages. The estimated CERs (carbon emission reductions) as a result are 77 000 MT (million tonnes).

Case study

NOx and environmental hazards

NO_x is the generic term for a group of highly reactive gases, all of which contain nitrogen and oxygen in varying amounts. Many of the NO_x gases are colourless and odourless. However, one common pollutant, NO_2 (nitrogen dioxide), along with particles in the air, can often be seen as a reddish-brown layer over many urban areas. NO_x is formed when fuel is burned at high temperatures, as in a combustion process. The primary sources of NO_x are motor vehicles, electric utilities, and other industrial, commercial, and residential sources that burn fuels.

NO_x is one of the main ingredients in the formation of ground-level ozone, which can trigger serious respiratory problems if it reacts to form nitrate particles, acid aerosols, as well as NO_2, which can also cause respiratory problems. NO_x contributes to the formation of acid rain and also nutrient overload that deteriorates water quality. It contributes to atmospheric particles that cause visibility impairment, most noticeable in national parks. NO_x reacts to form toxic chemicals and also leads to global warming.

Several technologies are available for reducing NO_x; the thermal $DeNO_x$ technology implemented at CPCL is explained below.

The thermal $DeNO_x$ process

Thermal $DeNO_x$ is a selective non-catalytic process for the reduction of NO_x emissions from stationary combustion sources. It is based on the gas phase homogenous reaction between NO (nitric oxide) and ammonia to produce nitrogen and water. For the process, anhydrous ammonia is injected into the flue gas by means of a carrier LP (low-pressure) steam at a location selected to provide the optimum combination of reaction temperature and residence time. The simplified reaction is as follows.

$$NO + NH_3 + O_2 + H_2 \rightarrow N_2 + H_2O$$

The thermal $DeNO_x$ facilities at CPCL, Manali have been developed by EMRE (Exxon Mobil Research and Engineering Company), based on a proprietary kinetic model, which solves 43 equations to identify the optimum reaction conditions for NO_x reduction.

Thermal $DeNO_x$ system description

Anhydrous ammonia is used as the reagent for the thermal $DeNO_x$ system. The system consists of a common supply of anhydrous ammonia for all heaters and boiler, along with associated control systems for the heaters and boiler.

One horizontal anhydrous ammonia drum is provided for storing anhydrous ammonia. One ammonia-unloading compressor has been provided for unloading ammonia from the road tankers to the storage drum.

Anhydrous ammonia is supplied to heaters and boiler from the ammonia storage drum. The anhydrous ammonia is vapourized in the heat exchanger with LP steam as the heating medium. LP steam is used as a carrier medium for injecting ammonia into firebox, which is taken from the respective units.

To achieve the desired NO_x reduction at lower loads, hydrogen injection is required, which is met by the new HGU (hydrogen generation unit). Hydrogen is supplied to all heaters through a common header with individual distribution and control systems for each heater. Typically, the initial NO_x levels of 300–325 PPM (parts per million) are reduced to 80–100 PPM, depending upon the feed case of the fuel oil.

The thermal $DeNO_x$ system has been designed to reduce NO_x emission from the CDU atmospheric heater, VDU (vacuum distillation unit) vacuum heater, OHCU (once through hydro cracker unit) fractionator feed heater, VBU (vis breaker unit) heater, NHT (naphtha hydro treater) reactor feed heater, NHT stripper reboiler, CRU (catalytic reforming unit) stabilizer reboiler, and boiler 4 in the refinery, based on the heat duty of the furnace and fuel quantity used.

CPCL has installed low-NO_x burners, and they are in operation in all the process heaters and furnaces of the DHDS and refinery III complex. Low-NO_x burners reduce NO_x emission at the source.

The $DeNO_x$ facility is implemented in India for the first time and CPCL has the privilege to commission the same. The $DeNO_x$ facility is an end-of-the-pipe treatment to minimize the overall NO_x emissions. The total cost of the $DeNO_x$ system is approximately Rs 110 million.

CASE STUDY

at a glance

- CPCL (Chennai Petroleum Corporation Ltd) is a group company of the Indian Oil Corporation Ltd.
- The company produces LPG (liquefied petroleum gas), motor spirit, superior kerosene, ATF (aviation turbine fuel), high-speed diesel, naphtha, bitumen, lube base stock, paraffin wax, fuel oil, hexane, and petrochemical feed stocks, and had sales of over Rs 211 billion for the year ending 31 March 2006.
- NO_x (oxides of nitrogen) is one of the main causes of the formation of ground-level ozone, which can trigger serious respiratory problems. It reacts to form nitrate particles, acid aerosols, as well as NO_2 (nitrogen dioxide), which also causes respiratory problems.
- Further problems resulting from NO_x include the formation of acid rain, deterioration of water quality due to nutrient overload, visibility impairment (most noticeable in national parks), and global warming.
- NO_x removal is achieved at CPCL by the use of low-NO_x burners and end-of-pipe treatment using $DeNO_x$ technologies.
- Thermal $DeNO_x$ is a selective non-catalytic process for the reduction of NO_x emission from stationary combustion sources. It is based on a gas phase homogenous reaction between NO (nitric oxide) and ammonia to produce nitrogen and water. For the process, anhydrous ammonia is injected into the flue gas by means of a carrier low-pressure steam at a location selected to provide the optimum combination of reaction temperature and residence time.
- With the $DeNO_x$ facility, the initial NO_x levels of 300–325 PPM (parts per million) reduced to 80-100 PPM (depending upon the feed quality of fuel oil); thus, NO_x emission to the atmosphere is reduced by three to four times.
- CPCL is the first company in India to implement the thermal DeNOx technology for the removal of NO_x from the flue gases, and has become a role model for other refineries and major industries.

CASE STUDY

12

Indian Farmers Fertiliser Cooperative Ltd

IMPLEMENTATION OF ENERGY SAVING SCHEME IN IFFCO AONLA AMMONIA-I

Background

IFFCO (Indian Farmers Fertiliser Cooperative Ltd) was established as the farmers' own initiative in November 1967 as a multi-state cooperative organization. It had broad objectives of augmenting fertilizer production, ensuring the availability and strengthening the distribution system of fertilizer, as well as educating, training, and guiding the farmers for improving the agricultural and rural economy.

IFFCO comprises three ammonia–urea complexes and two mixed fertilizer plants. The company's ammonia–urea complexes are situated at Kalol in Gujarat, Phulpur in Allahabad, and Aonla in Uttar Pradesh, while the two mixed fertilizer plants are situated at Kandla in Gujarat and Paradeep in Orissa. IFFCO's flagship Aonla Unit (Figure 1) is a leading manufacturer of fertilizer-grade urea, producing 1.74 MT (million tonnes) in 2005/06. It achieved an annual turnover of approximately Rs 99.43 billion for the same period.

Environmental profile

Certifications and awards

IFFCO has implemented an SHE (safety, health, and environment) policy at the corporate level as well as an environmental policy, and has adopted EMS (environmental management system) since its design stage. In August 2000, IFFCO Aonla aligned its EMS as per the requirements of ISO 14001: 1996 and upgraded to the ISO 14001: 2004 standards in April 2006. Apart from these, IFFCO Aonla Unit is also an ISO 9001: 2000 and OHSAS 18001: 1999 certified unit.

It also has a separate environment and pollution control cell with an environmental laboratory, where the quality of storm water is checked on

Figure 1
IFFCO Aonla Unit

a daily basis, urea dust from urea prill towers and other ambient air parameters are continually monitored, and the quality of liquid effluent is tested both before and after treatment from different locations in the plant. The standard of these environmental parameters is also monitored by the UPPCB (Uttar Pradesh Pollution Control Board), Bareilly.

As a result of IFFCO Aonla's various efforts in environment improvement, energy conservation, and safety and health practices, the company has won many awards. They received the Indo-German Greentech Environment Excellence Award in 1999/2000 and in 2000/01. They also received the NSCI (National Safety Council of India) Safety Award–2002 for excellence in the safety management and the National Conservation Award–2003 from the Ministry of Power, Government of India, among others.

Air pollution

The major air pollutants emitted by the plant are SPM (suspended particulate matter), SO_2 (sulphur dioxide), NO_x (oxides of nitrogen), and fugitive emissions. The source of SPM is the urea dust that is emitted from the prilling towers in the urea plants and product handling plant. The height of the prilling tower is kept at 96 metres in Aonla I and 104 metres in Aonla II, and the prilling bucket is designed such that urea dust remains well below 50 mg/Nm^3, the standard limit of the UPPCB. The average value of urea dust in the IFFCO Aonla for 2005/06 was 39.10 mg/Nm^3. Both the urea and product handling plants have dust extraction and recovery systems where the dust is collected, dissolved, and reutilized in the processes in the urea plant. The fuels used by IFFCO Aonla are clean fuels and trace between 1 and 30 PPM (parts per million) of SO_2, well below the UPPCB limit of 100 PPM. For the effective and complete combustion of fuels, sufficient oxygen is maintained in the furnaces, and for continuous monitoring of oxygen, CO (carbon monoxide), and NO_x, analysers have been installed. NO_x concentration is kept in the range of 20–60 PPM, well below the standard limit of 100 PPM set by UPPCB. Finally, proper care has been taken in the design stage to control the fugitive emissions/odour nuisance. Pressurized air control rooms and lobbies in the plant have been constructed, and sufficient numbers of compressed air-breathing apparatus, gas masks, emergency air-breathing apparatus, and eye-wash showers have been provided at respective locations in the plant.

Noise pollution

In order to reduce noise pollution, silencers and enclosures have been provided. The silencers have been provided for all the vents in order to prevent noise pollution.

Water pollution

The water pollutants emitted from the IFFCO Aonla Unit are pH, BOD (biochemical oxygen demand), COD (chemical oxygen demand), TSS (total suspended solids), TDS (total dissolved solids), oil and grease, and thermal discharge. The pH of the effluent discharged from the battery unit of the factory is maintained within the UPPCB limit of 5.5–8.5, and measures have been taken to further decrease this. In order to tackle BOD, the plant and canteen have separate septic tanks that are dosed with bleaching powder, and there is an oxidation pond for the treatment of sewage water. Also, the COD value in the treated effluent remains in the range of 15–100 mg/litre, against the UPPBC standard of 250 mg/litre. In order to reduce TSS and TDS, the raw water utilized is drawn from 250–300-metre-deep bore wells and the cooling water treatment is based on a non-chromate organo-phosphonate-based bio-degradable corrosion inhibitor. Disc oil separators have been provided in the ammonia and urea plants to recover the oil and grease from the oily waste water. The oil is skimmed by the disc oil separator and collected in drums that are sold to recyclers, and the remaining water is sent to the ETP (effluent treatment plant) for further treatment. No thermal discharge is allowed to mix in the effluent going to the ETP.

Hazardous waste management

The hazardous waste generated by IFFCO is chromium sludge, spent catalyst, spent oil, and acid–alkali waste. In order to reduce chromium sludge, the use of chromate-based corrosion inhibitors has been phased out and eco-friendly organo-phosphonate-based corrosion inhibitors have been used instead. The spent catalyst and spent oil are stored in seal covered drums as per the CPCB (Central Pollution Control Board) guidelines. The acid–alkali waste water is collected in the neutralization pit and corrected to the UPPCB standard of pH 6.5–8.0.

Case study

Introduction of the energy-saving scheme

The ammonia–urea complex at IFFCO's Aonla Unit comprises two phases, Aonla Stream I and Aonla Stream II. The total capacity of both these streams is 1 003 200 MTPA (million tonnes per annum) of ammonia and 1 729 200 MTPA of urea. With effect from April 2003, the Government of India New Group Pricing Scheme for the Urea Fertilizer Industry laid stringent energy norms and made it imperative to reduce the cost of production by reducing energy consumption.

A two-phase techno-economically feasible energy-saving scheme was introduced at the unit. The first phase of the scheme, which is the focus of this case study, covered the modification of front-end of the ammonia plant by modifying the CO_2 (carbon dioxide) removal system along with LTS (low temperature shift) guard with BFW (boiler feed water) pre-heater. The second phase covered the modification of back-end of ammonia module in waste heat recovery section of primary reformer of Ammonia II.

The energy-saving scheme does not generate any liquid or gaseous effluent; instead, it reduces the consumption of natural resources being used as fuel, feed, and utilities, and reduces the emission of CO_2.

Conversion of single-stage GV system to two-stage GV system for CO_2 removal in ammonia plant-I

The original system

In the original system, the process gas containing CO_2 enters at the bottom of the GV (Giammarco Vetroke) absorber, and the GV solution enters the absorber from the top and absorbs the CO_2 from the process gas while flowing down through the packed beds. The CO_2-free gas then comes out of the top of the absorber and is sent down for further processing. The rich GV solution containing the CO_2 is sent to a regenerator operating at about 1 kg/cm^2 g. Here, the dissolved CO_2 is separated from the solution by flashing and by the stripping action of rising steam and vapours. The lean GV solution collects at the bottom of the regenerator and goes to a flash drum operating at 0.6 kg/cm^2 g, where the dissolved CO_2 is separated once more. The separated CO_2 is fed to the regenerator from the flash drum with the help of the LP (low pressure) steam ejectors. The lean GV solution is pumped from the flash drum to the GV absorber, while the separated CO_2 coming out from the regenerator top is cooled and sent to the urea plant.

The modified system

The main modifications of this process by the energy-saving scheme are absorption by lean and semi-lean solutions, and by high- and low-pressure stripping. These changes have resulted in better absorption of CO_2 in the absorber and lower energy consumption for regeneration of the solution in regenerators.

In the modified system, the process gas containing CO_2 enters at the bottom of the GV absorber, while the lean GV solution enters the absorber at the top and absorbs CO_2 from the process gas while flowing down through the packed beds. The free CO_2 comes out from the top of the absorber and is sent for further processing. The semi-lean solution enters at the middle of the same absorber and absorbs CO_2 from the process gas. Lean solution is pumped to the top of the absorber to absorb the remaining CO_2 in the gas. When the GV solution reaches the bottom of the absorber, it becomes rich in absorbed CO_2.

The rich GV solution containing CO_2 is sent to the HP regenerator operating at about 1 kg/cm^2 g. Here, part of the dissolved CO_2 is separated from the solution by flashing. The first part of the GV solution is sent from the HP regenerator top take off tray to the LP regenerator top operating at about 1 kg/cm^2 g. The second part is sent from the HP (high pressure) regenerator semi-lean draw-off pan to the LP regenerator semi-lean solution sump. The remaining GV solution in the HP regenerator is reboiled in the GV reboilers. The vapours from the reboilers re-enter the HP regenerator in order to strip the GV solution. Solution from reboilers is sent to HP regenerator bottom. Lean solution from HP regenerator bottom is sent to LP regenerator bottom. The semi-lean solution from the LP regenerator draw-off pan is pumped to the GV absorber and enters above the top of the second bed for absorption of CO_2. Lean solution from the bottom of the LP regenerator is cooled in a heat exchanger with DM water and is pumped to the GV absorber top. CO_2 from the LP regenerator is boosted up to HP regenerator pressure with a CO_2 blower and sent to the urea plant along with the CO_2 from the HP regenerator.

Benefits of the new system

The main benefits of these modifications are, first, the reduction of CO_2 slip by about 600 PPM, which resulted in higher availability of CO_2 for urea production and a decrease in hydrogen consumption in the methanation section. Second, it results in a reduction in the LP steam consumption in the CO_2 removal system from 38 MT/h to 15 MT/h.

LTS guard reactor and BFW pre-heater

The second initiative is with the LTS guard reactor (Figure 2) and BFW pre-heater. In the ammonia–urea manufacturing complex, ammonia is manufactured by steam reforming of natural gas or naphtha. During the process, CO is formed in the gaseous mixture that is converted into CO_2 and hydrogen.

The original system

Prior to modification, the reformed gas from the reforming section flowed to the HT shift converter after cooling in and HP waste heat boiler from 988 °C to 380 °C. The CO content of the process gas was reduced from 12.96% to 3.46% in the HT shift converter through shift reaction, which takes place in the reactor in the presence of an iron-chromia catalyst. The temperature of the process gas is then reduced from 444 °C to 210 °C by

Figure 2
LTS guard reactor

heat recovery in a waste heat boiler and BFW pre-heater. The CO content of 3.46% in the process gas is then further reduced to 0.37% in the LTS converter through a shift reaction, which takes place in the reactor in the presence of chromium-based and copper-zinc catalysts.

The modified system

The modifications that IFFCO made to this process entailed installing a new LTS guard reactor to reduce the CO slippage from the shift conversion section. The CO slip is then lowered down to 0.10%–0.16%, instead of 0.37% without the LTS guard reactor. The lower CO slip results in additional ammonia production due to reduction in the consumption of hydrogen in the methanator. A new BFW pre-heater was also installed in order to control the process gas inlet temperature to the LTS converter.

Environmental benefits of phase 1 of scheme

The implementation of phase 1 of the energy-saving project at IFFCO Aonla has resulted in a reduction of specific energy consumption per tonne of ammonia, and energy saving amounting to 0.258 Gcal (gigacalories)/MT of ammonia in 2005/06. Further, it has saved a substantial amount of natural gas, which is a precious raw material, and ultimately reduced CO_2 emissions to the atmosphere. For the same year, about 26 400 MT of CO_2 gas emission was reduced. Monetary savings from the scheme are outlined in Table 1.

TABLE 1
COST–BENEFIT ANALYSIS

Description	*Ammonia-I*
Energy saving (Gcal/MT of ammonia)	0.24
Ammonia production (MT)	478 692.00
Investment (in Rs million)	310.40
Total energy saving (Gcal)	113 928.70
Cost of energy per Gcal	2475.55
Total monetary saving (in Rs million)	282.04
Payback period (months)	13.21
Carbon dioxide saving (MT)	26 397.58

MT - million tonnes; Gcal - gigacalories

CASE STUDY

at a glance

- IFFCO, established in November 1967, comprises three ammonia–urea complexes and two mixed fertilizer plants.
- IFFCO's flagship Aonla Unit is a leading manufacturer of fertilizer-grade urea, producing 1.74 MT (million tonnes) in 2005/06, and achieving an annual turnover of approximately Rs 99.43 billion for the same period.
- To upgrade existing plant technology and reduce energy consumption, a two-phase energy-saving project was developed.
- The case study focuses on the first phase of the scheme, which was completed in June 2005, and entails the revamping of the CO_2 (carbon dioxide) removal system to a two-stage GV (Giammarco Vetroke) process and installation of a new LTS (low temperature shift) guard reactor and BFW (boiler feed water) pre-heater.
- The project resulted in an energy saving of 0.24 Gcal/MT of ammonia in 2005/06 and reduced CO_2 emissions by 26 400 MT.
- Total monetary saving of Rs 282 million was achieved.
- The payback period of the project is 13.21 months.

CASE STUDY

13

Indian Oil Corporation Ltd

Resource conservation and elimination of hazardous chemicals through change of solvent in amine treating units for desulphurization of refinery gases

Background

The Mathura Refinery of IOC (Indian Oil Corporation Ltd) produces LPG (liquefied petroleum gas), propylene, naphtha, MS (motor spirit), ATF (aviation turbine fuel), SK (superior kerosene), HSD (high-speed diesel), FO (furnace oil), HPS (heavy petroleum stock)/RFO (residual fuel oil), bitumen, and sulphur. IOC's annual turnover for the year 2005/06 was in excess of Rs 1752 billion.

Environmental profile

IOC's environmental objectives have evolved as a part of its EMS (environmental management system). The main objectives of the Mathura Refinery are to maintain its overall SO_2 (sulphur dioxide) emission within the prescribed limit of 450 kg/h, reduce the energy consumption level progressively, reduce fugitive emission of hydrocarbons, reduce consumption of raw water, plant trees in the Mathura region, properly dispose oily sludge in order to protect the soil and groundwater from contamination, develop environmental awareness amongst all refinery personnel, and educate 100 truck drivers per year on the hazards in the transportation of petroleum products.

The main water pollutants emitted by IOC are pH, phenol, oil and grease, sulphide, BOD (biochemical oxygen demand), and TSS (total suspended solids). All these pollutants are kept within the limits of their prescribed standards.

The two categories of hazardous waste generated by IOC are oily sludge and spent catalyst. The sludge is generated from the storage tank bottoms and ETP (effluent treatment plant) basins. It is minimized by hot gas oil circulation before the cleaning of the crude oil tanks and the operation of the melting pit. These operations recover oil from the sludge, and the recovered oil is reprocessed while the residual sludge is disposed off into a PVC (polyvinyl chloride)-lined pit that has a proper leachate collection facility. The spent catalysts are generated only during replacement, and are sold to a party that has valid registration with the CPCB (Central Pollution Control Board). Biological sludge generated from the ASP (activated sludge process) in the ETP is not hazardous in nature; after dewatering and stabilization, it is used as bio-manure, thereby avoiding the use of chemical fertilizers.

Case study

Background

Crude oil contains undesirable sulphurous compounds, which are distributed in products during processing. These products are treated to lower the sulphur content by hydro-treating, which generates sour gases, mainly constituting H_2S (hydrogen sulphide) that can be removed through the process of amine treating.

Sulphur management to meet stringent environmental standards

More stringent environmental standards for the emission of SO_2, along with low sulphur specifications for petroleum products, have resulted in making sulphur management critical in refinery operations. The innovations and advancements of sulphur management facilities have become more prudential with increased processing of sour crude oils on preferential basis, demand for clean fuels, and other environmental needs.

The process of desulphurization

Sulphur management in a refinery consists of three basic processes. The first process is to remove H_2S from refinery gases and LPG streams at the ATU (amine treating unit). The second process is the separation of H_2S from the solvent at the ARU (amine regeneration unit). The final process is at the SRU (sulphur recovery unit) where H_2S is converted to elemental sulphur.

Amine treating unit

The ATU is an integral part of the desulphurization process (Figure 1). This is a gas/liquid system in which alkaline amine solvent scrubs the H_2S and is passed through a flash drum to remove hydrocarbons. The settler also serves as a feed drum for the ARU. In the ARU, the H_2S is separated from the 'rich amine' solvent and the 'lean amine' solvent is then returned to the absorber tower along with the addition of fresh make-up solvent. The H_2S produced from the ARU is fed into the SRU for production of elemental sulphur.

Solvent selection

The three widely used alkaline amine solvents are MEA (monethanolamine), DEA (diethanolamine), and MDEA (methyldiethanolamine). MEA is generally not preferred, while DEA is selected where relatively small amounts of H_2S are present. Of the three solvents, MDEA is highly selective towards H_2S and has the highest H_2S absorption capacity. Therefore, if the current solvent is DEA, more capacity can be created in the same equipment by switching over to MDEA, thus, less solvent would be required for the same amount of sour gases.

Resource conservation through solvent selection

The superiority of MDEA over DEA is well documented and many refineries across the world have switched over to MDEA. Despite MDEA being more expensive, substantial monetary savings can be achieved through resource conservation and energy-saving. As MDEA is the most acceptable solvent in terms of selectivity and absorption capacity for H_2S, less solvent is required for the same amount of sour gases. This also means that there is less consumption of steam in the amine regeneration unit for the regeneration of the solvent. Besides monetary gains, the less hazardous nature of MDEA is an added advantage.

By switching to MDEA, the Mathura Refinery has benefited as first, the treating capacity of the existing amine treating units have been enhanced by 30%; second, MDEA has a higher selectivity for H_2S and maximum 'rich amine' loading of 0.45 moles of H_2S/mole of amine can be achieved against the 0.33 moles of H_2S/mole of amine for DEA; third, there is less stripping steam required to regenerate MDEA and this has resulted in 37% steam reduction in the re-boiler of the ARU; fourth, MDEA – being in liquid form – is easier to handle in comparison to DEA,

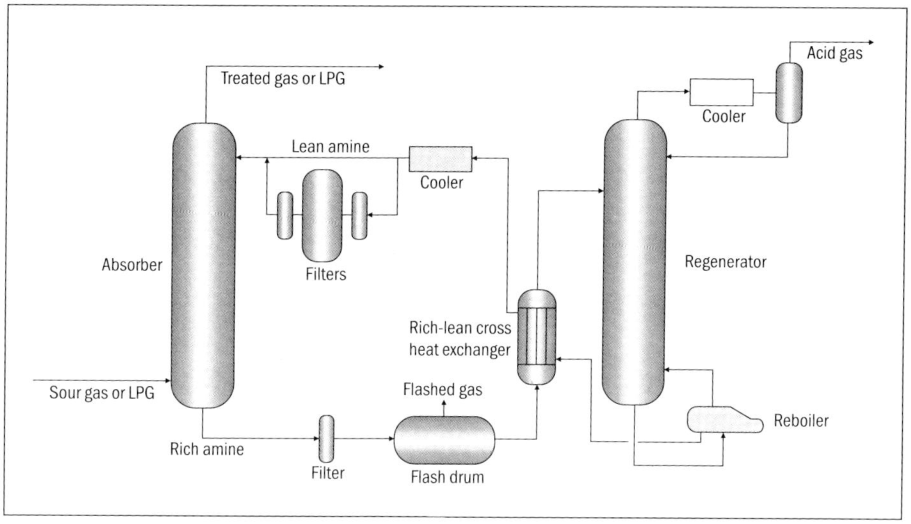

Figure 1
Desulphurization process

which is supplied in a semi-solid form and has to be melted by steam before being taken into the system; and, finally, MDEA is less hazardous in comparison to DEA.

Cost benefit

The cost of MDEA is five per cent higher than that of DEA and, therefore, the incremental expenditure on amine consumption due to the switch-over from DEA to MDEA works out to be Rs 330 000 per year. However, the net financial savings for the refinery amounts to Rs 50 million per year.

CASE STUDY

at a glance

- The Mathura Refinery of IOC (Indian Oil Corporation Ltd) produces LPG (liquefied petroleum gas), propylene, naphtha, MS (motor spirit), ATF (aviation turbine fuel), SK (superior kerosene), HSD (high-speed diesel), FO (furnace oil), HPS (heavy petroleum stock)/RFO (residual fuel oil), bitumen, and sulphur.
- The ATU (amine treating unit) is an integral part of the desulphurization process.
- The three widely used alkaline amine solvents are MEA (monoethanolamine), DEA (diethanolamine), and MDEA (methyldiethanolamine).
- Earlier, DEA was being used at Mathura Refinery to scrub H_2S (hydrogen sulphide) from refinery gases in various ATUs; however, the refinery recently switched to MDEA through various in-house efforts, resulting in immense benefits to the company.
- MDEA is the most acceptable solvent in terms of selectivity and absorption capacity for H_2S; thus, less solvent is required for the same amount of gases. This also implies less consumption of steam in the ARU (amine regeneration unit) for regeneration of the solvent.
- MDEA is a far less hazardous solvent than DEA.
- The switch-over has enhanced the treating capacity of the existing ATUs by 30%.
- MDEA has a higher selectivity for H_2S and maximum 'rich amine' loading of 0.45 moles of H_2S/mole of amine can be achieved, against 0.33 moles/mole for DEA.
- Less stripping steam is required to regenerate MDEA.
- The switch-over has also resulted in monetary gains for IOC of approximately Rs 50 million each year.

CASE STUDY

14

Indo Gulf Fetilisers Ltd

Reduction in CO_2 emission through energy efficiency improvement in CO_2 removal system in ammonia project

Background

IGF (Indo Gulf Fertilisers), a unit of ABNL (Adity Birla Nuvo Ltd), produces approximately 864 000 MT (metric tonnes) of urea each year. The plant is located in the Sultanpur district, Uttar Pradesh, and was commissioned in October 1988. Net profit for 2005/06 was Rs 1869.3 million.

Environmental profile

IGF has an environment policy and department at both the corporate and facility levels. The facility also has environmental risk management and disaster management plans, to reduce the impact of pollutants on the environment and mitigate any type of emergency that may arise, respectively.

The unit takes several initiatives under its environment management programme, including improving awareness on environmental issues amongst employees, vendors, and sub-surface contractors; planting at least 1000 trees each year; maximizing the utilization of treated effluent in the green belt; and ensuring that water consumption is maintained at a level below 5 m^3(cubic metres)/MT of urea. In 2005/06, green belt development outside the complex included the plantation of 9000 jatropha plants.

Being a nitrogenous fertilizer plant, the major air pollutants emitted include SPM (suspended particulate matter) and ammonia from the prilling tower, and SPM, RSPM (respirable suspended particulate matter), and NO_2 (nitrogen oxide) from the boilers, gas turbine, and primary reformer stack. To ensure that all pollutants are within the stipulated norms of the PCB (pollution control board), IGF takes several measures. Environment-friendly technology is used in the manufacture of urea, such as a natural-gas-fired primary reformer, boilers, and gas turbines that have less sulphur content to control SO_2 emission. Other efforts include provision of sufficient stack height for flue gases, adoption of the latest technology for complete automation of the system, and online and offline monitoring of critical pollutants. To control fugitive emission that occurs in the bagging plant, two wet de-dusting systems have been provided that suck dust from emission points and send it to the urea plant for reprocessing. Cyclone separators have also been installed at transfer points; dust collected at these points is further dissolved in urea solution for reprocessing in the urea plant.

The plant uses the 'at source treatment and recycle to process' philosophy to govern most of its effluent streams. The three-stage treatment system – comprising primary, secondary, and tertiary treatment stages – ensures that water pollutants are kept at their minimum.

The two major categories of hazardous waste generated are spent catalyst and used oil. Both are stored in leak-proof steel drums of adequate capacity and sold to PCB-authorized parties or used for application on the surface of pipes lying in the yard to protect those that come in direct contact with the atmosphere from corrosion. Activated carbon, an industrial waste, is used for the purification of the GV (Giammarco Vetrocoke) solution in the ammonia plant for removing organic matter from the water. Used batteries from the UPS (uninterrupted power supply) are sold to registered recyclers, and ruptured HDPE (high density polyethylene) bags are collected and sold to parties for recycling or reuse.

Further, measures are taken towards water conservation, including recycling of condensate boiler feed water, reuse of treated effluent water for irrigation in the green belt, reuse of boiler blow-down water as cooling tower make-up water, as well as implementation of rainwater harvesting schemes on the complex. Figure 1 shows the reduction in process water consumption from 2002/03 to 2005/06.

IGF has registered its project to reduce CO_2 (carbon dioxide) emissions in its ammonia plant under the CDM (clean development mechanism), the details of which are provided in the case study.

Case study

IGF's CO_2 removal system in its ammonia plant, in place to abate GHGs (greenhouse gases) that contribute to global warming, was identified as an area for improvement due to its high energy consumption. For improving the energy efficiency of the CO_2 removal system, an innovative approach was adopted based on the basic principles of chemical engineering. The key challenge was to find a cost-effective solution that could ensure safe and sustained plant operations. The CO_2 removal system, modified for low-energy consumption with the help of GV, was commissioned in August 2003 and has been running successfully since then. The project was taken up as a CDM project under UNFCCC (United Nations Framework Convention on Climate Change).

After the preliminary study carried out by GV, the technology supplier, and IGF, it was evident that the improvement in energy consumption would be possible by converting the original system into a lean (5% bicarbonate) or semi-lean (10% bicarbonate) system. A first option entailed converting the complete CO_2 removal system into a standard low energy GV process; however, the cost for implementing this was proving too high, as it required a bigger LP (low pressure) tower. The second option was to increase the height of the existing LP tower by five metres, to the height of 46 metres, which also proved unfeasible as the existing foundation of the LP tower would not be able to take additional liquid and wind load, and would require a long shut-down time.

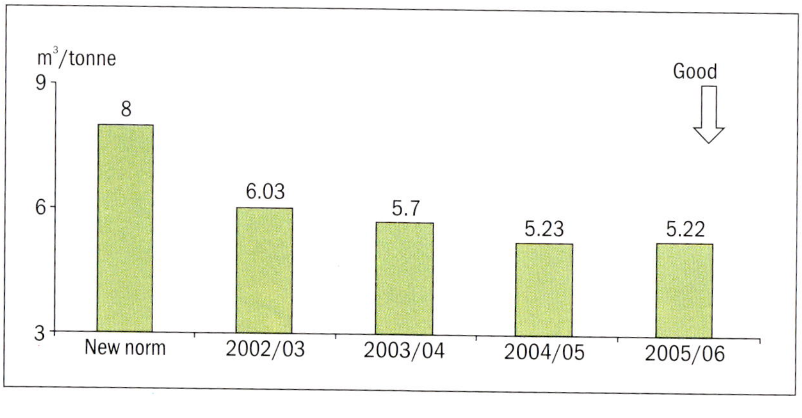

Figure 1
Process water consumption

An innovative approach was, hence, designed whereby the tower size would be increased with an additional smaller vessel, which would, in turn, reduce the project cost. The process was converted into a lean/semi-lean type and the final scheme had the following features.

- A small flash vessel was installed, operating at 0.03 MPa (mega pascal).
- Provision of a new take-off tray in the HP (high pressure) tower was made to extract semi-regenerated solution and send it to the existing flash vessel. The normal existing path to the LP tower bottom was then followed. This solution is used as semi-lean solution and is fed into the middle of the absorber.
- Flash steam or CO_2 from the intermediate flash vessel is connected to the bottom of the new take-off tray in the HP tower, instead of the HP tower bottom.
- The solution flow from the HP to the LP tower, which was earlier 200 m^3/hour, is increased to 350 m^3/hour.
- The purer solution at the bottom of the HP tower is flashed in a new flash vessel. After flashing, this regenerated solution is used as lean solution and flashed steam is utilized in the existing LP tower for regeneration. The vapours generated by flashing of this solution in the new flash vessel helps to supply additional heat for regeneration in the LP tower. The solution from the bottom of the new flash vessel becomes extra pure (lean solution) and is sent to the absorber by a new lean solution pump with cooling in a demineralized water pre-heater, instead of air fans.

The modified CO_2 removal system, based on the underlying principle of converting it into a lean/semi-lean system by shifting the load from the HP to the LP tower, resulting in a reduction of steam consumption in the HP regenerator, is depicted in Figure 2. As shown in the figure, the heat, which was being dumped into the atmosphere while cooling the GV solution through air fan coolers, is now taken in the demineralized water unit.

As a result of the project, implemented in August 2003, steam consumption reduced from 32 tonnes/hour to around 15 tonnes/hour, and CO_2 slip reduced from 700 PPM (parts per million) to about 380 PPM. The reduction in steam consumption has further resulted in

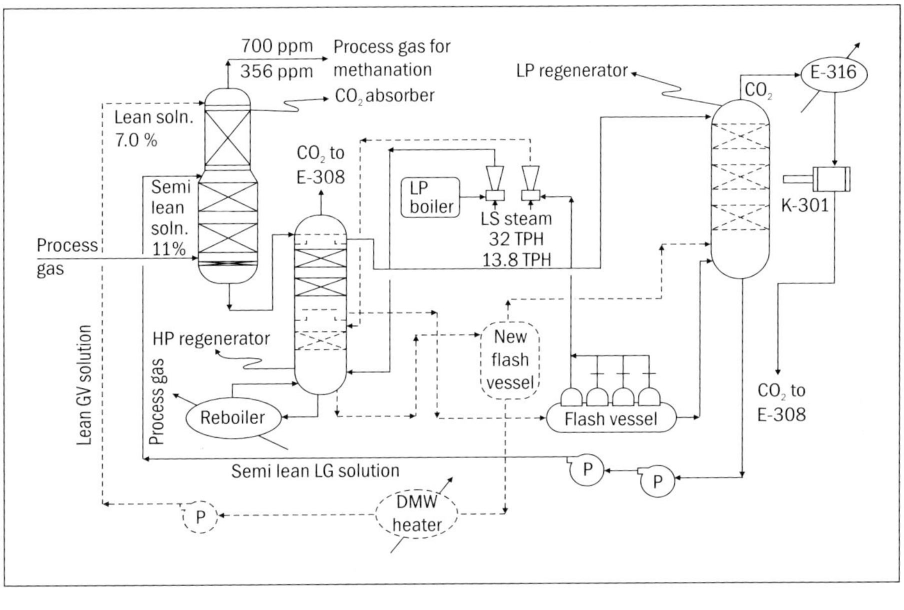

Figure 2
Modified CO_2 removal system

reduced natural gas and naphtha firing in service boilers, thus leading to reduced emissions of GHGs from the service boiler stack. The lowered energy cost also enhances the productivity of the plant.

Moreover, the project resulted in cost-benefits for IGF. The total cost of investment in the project was estimated at Rs 40 million. Due to the saving of about 17 tonnes/hour around steam, a saving of Rs 20 million was established, with a payback period of two years. In addition, the company benefits by Rs 10 million per year from the CDM through the sale of 60 239 MT CERs (certified emission reduction) to Barclays Bank Plc. IGF is the first fertilizer company in India to have successfully secured a CER certificate on the reduction of CO_2 emission from the UNFCCC.

CASE STUDY at a glance

- IGF (Indo Gulf Fertilisers), a unit of ABNL (Adity Birla Nuvo Ltd), produces approximately 864 000 MT (metric tonnes) of urea each year.
- The case study was implemented at the company's plant in the Sultanpur district, Uttar Pradesh.
- The modification of the CO_2 (carbon dioxide) removal system at the unit's ammonia plant was taken up as a CDM (clean development mechanism) project.
- The CO_2 removal system was modified based on the principle of converting the original system into a lean (5% bicarbonate) or semi-lean (10% bicarbonate) system.
- The modification was achieved by a number of features that shifted the load from the HP (high pressure) to the LP (low pressure) tower, resulting in a reduction of steam consumption in the HP regenerator, and by cooling GV/lean solution with demineralized water, instead of using air fans and dumping heat into the atmosphere. As a result of the project, steam consumption reduced from 32 tonnes/hour to about 15 tonnes/hour, and CO_2 slip reduced from 700 PPM (parts per million) to about 380 PPM.
- The reduction in steam consumption has further resulted in reduced natural gas and naphtha firing in service boilers, thus leading to reduced emissions of GHGs (greenhouse gases) from the service boiler stack.
- The essential benefits to the environment by the project are the reduction in consumption of natural resources (that is, natural gas and naphtha), and the reduction in all associated emissions due to reduced combustion of natural gas and naptha, including CO_2.
- The project has resulted in yearly savings of Rs 20 million for IGF, with a payback period of two years. In addition, it receives Rs 10 million per year from the sale of 60 239 MT CERs (certified emission reduction) under the CDM.

CASE STUDY

15

Numaligarh Refinery Ltd

Installation of stepless control system in make-up gas compressor in hydrocracker unit

Background

NRL (Numaligarh Refinery Ltd) is situated in Numaligarh, in the Golaghat district of Assam. The range of products manufactured at the facility include LPG (liquefied petroleum gas), naphtha, MS (motor spirit), ATF (aviation turbine fuel), SKO (superior kerosene oil), HSD (high-speed diesel), RPC (raw petroleum coke), CPC (calcined petroleum coke), and sulphur. The annual turnover of the company for 2005/06 was Rs 58 203 million.

Environmental profile

Since its inception, NRL has been pursuing a focused programme towards environment management by formulating a comprehensive policy towards its commitment for the protection, preservation, and development of the environment. NRL thus strictly adheres to the requirements of its EMS (environment management system), conforming to the internationally recognized standard ISO 14001: 2004, along with ISO 9001: 2000 for QMS (quality management system) and OHSAS (occupational health and safety management system) 18001: 1999. NRL was also awarded ISRS (International Safety Rating System) Levels 8 and 9.

With CSR (corporate social responsibility) being one of the principal objectives of the company, NRL is committed to improving the quality of life in society in general and of the local community living within a radius of 10 km of the refinery. Community development is facilitated through encouraging income generation activities; providing facilities for education and community health care; developing infrastructure; and promoting art, sports, literature, and culture.

Under the CDM (clean development mechanism), NRL has undertaken captive power generation by recovery and utilization of the waste energy (thermal and pressure) of HP (high pressure) steam. The project will reduce the release of about 34 000 TPA (tonnes per annum) of CO_2 (carbon dioxide) gas and other greenhouse gases to the environment.

NRL has taken several steps towards the minimization of environmental pollution. Major air pollutants emitted include SPM (suspended particulate matter), SO_2 (sulphur dioxide), NO_x (oxides of nitrogen), fugitive emissions, odour, and noise. Despite processing low sulphur Assam crude, NRL has installed an SRU (sulphur recovery unit), a sour water stripping unit, and sour gas sweetening unit. A non-illuminating ground flare, the first of its kind in the refining industry, has also been installed along with low-NO_x burners. Facilities are provided for continuous monitoring of SO_x, NO_x, etc. and the minimum height of the stacks for furnaces has been kept at 60 metres. NRL has adopted state-of-the-art energy-efficient technology, high-efficiency furnaces with glass air pre-heaters, captive cogeneration power plant using heat recovery system, maximization of waste heat recovery, installation of power recovery turbine in hydrocrakcer unit, etc.

In terms of water pollution, adequate segregation, collection, and centralized treatment facilities for waste water have been provided to meet the standards laid down under MINAS (Minimal National Standards) and by the pollution control board. A modern ETP (effluent treatment plant) with tertiary treatment facilities using latest technology with a three-stage oil recovery system has been installed. Moreover, utmost care is taken for the minimization of waste water generation and conservation of water resources.

Major categories of hazardous waste generated by the refinery are slop oil, oily sludge, and spent catalysts. Slop oil from process units is drained to the CBD (closed blow-down) system, from where oil is directly transferred to wet-slop tanks in the offsite area. Once water mixed with the oil settles, it is drained to the ETP for treatment. After the draining of water, dry-slop oil is transferred to the dry-slop oil tanks for reprocessing. The main process waste streams, the tank bottom drains, desalter effluent, and marketing terminal effluent are routed to the ETP through the OWS (oily water sewer system). After processing, oily sludge generated in the ETP is converted to an oil-free cake and is disposed off in a secured landfill, as total oil is recovered from the sludge and recycled back to the slop-oil system for reprocessing. Spent catalysts are either disposed off for metal recovery wherever possible, or disposed off in the secured landfill with leachate recovery and treatment system. NRL has approved environment-friendly disposal procedures for solid wastes other than ETP and tank bottom sludge such as metallic wastes; incinerable wastes such as paper, cotton, and cloth; and non-incinerable wastes such as fused bulbs, tube-lights, and batteries.

Case study

The stepless control system

Prior to the installation of the stepless control system (Figures 1a and 1b) at the make-up gas compressor in the HCU (hydrocracker unit), hydrogen could be supplied from the compressor in steps of 50%, 75%, or 100% only. The reciprocating compressors required to supply hydrogen continuously at a pressure of 187.7 kg/cm^2 (square metres). At partial plant load, the hydrogen compressed in excess was spilled back, which accounts for power

Figure 1a
Stepless control system

Figure 1b
Stepless control system

as well as fuel loss. With the stepless control system, feed is supplied in varying amounts in the range of 0%–100%, as per the plant requirements. The spill-back flow thus becomes zero, and energy is saved at variable plant load.

The stepless control system works on the principle of reverse flow regulation in the compressor that saves energy as compared to the bypass control. Hydraulically actuated unloaders keep the suction valves open during part of the compression cycle. Thus, part of the gas taken into the cylinder during the suction cycle is pushed back into the suction plenum. In this way, the gas volume per working stroke can be controlled. This system saves compression power at part load, since the energy consumption of a compressor is proportional to the quantity of gas compressed per compression cycle.

Energy recovery

Energy recovery by installing the stepless control system in NRL is the first of its kind in India. Prior to the introduction of the stepless control system, daily energy consumed at an average HCU load of 84.7% was 124 158 kWh (kilo-watt hours). However, after installation of the stepless control system, energy consumed at the same HCU load was 105 840 kWh. Thus, there was a daily power saving of 0.76 MW (megawatts).

Cost benefit analysis

NRL saved 3 MWh (mega-watt-hours) of power during 2006. Given the cost of power at Rs 3.3/kWh, the refinery saved Rs 10.3 million. Expenditure for the implementation of the project included Rs 11 million for installation of the stepless control system, Rs 726 000 per annum in depreciation (6.6% per annum assuming 15-year life of system), Rs 550 000 per annum in maintenance (5% cost of the system per annum), and Rs 1.1 million per annum in cost on interest on investment (10%), totalling over Rs 2.3 million. The recovery period for the system is approximately 16.6 months.

CASE STUDY

at a glance

- NRL (Numaligarh Refinery Ltd) is located in the Golaghat district of Assam, and had an annual turnover of Rs 58 203 million in 2005/06.
- NRL installed a stepless control system in the make-up gas compressor in the HCU (hydrockracker unit) in order to save energy.
- This system allowed hydrogen to be supplied in varying amounts in the range of 0%–100% (as opposed to the previous system, wherein feed could be supplied from the compressor in steps of 50%, 75%, or 100% only), as per plant requirements.
- The installation of the system thus resulted in zero spill-back flow and a subsequent saving of energy.
- During 2006, NRL saved 3 MWh (megawatt-hours) of power, which reflected in a cost saving of Rs 10.3 million.
- The recovery period of the system is approximately 16.6 months.
- The stepless control system at NRL is the first of its kind in India.

CASE STUDY

16

Oil and Natural Gas Corporation Ltd

FLARE GAS REDUCTION

Background

ONGC (Oil and Natural Gas Corporation) is one of the largest players in the oil and natural gas industry in the country. The ONGC Group – comprising of ONGC, wholly-owned subsidiary OVL (ONGC Videsh Ltd), and MRLP (Mangalore Refinery and Petrochemicals Ltd) – produced 1.11 million barrels of oil per day in 2005/06. The ONGC Group turnover for the same period was Rs 864.14 billion.

The company's GGS (group gathering station) is located in Narimanam, Nagapattinam district, Tamil Nadu, and falls under the umbrella of the Cauvery Asset, based in Karaikal, Puducherry. GGS-Narimanam (Figure 1) registers an annual production of 72 250 MT (million tonnes) of crude oil and 26.031 MSCM (million standard cubic metres) of natural gas. The Cauvery Asset, per se, records an annual turnover of Rs 10.82 billion in 2005/06.

Environmental profile

ONGC has obtained quality, health, safety, and environment certification from ISO 9001, ISO 14001, and OHSAS 18001. In addition, initiatives related to sustainable development have been taken. This includes a CDM (clean development mechanism) project for zero flare status at the GCS (gas collecting station), at Kuthalam, Nagapattinam, Tamil Nadu, which was submitted to the MoEF (Ministry of Environment and Forests) in 2005/06. Also, Halon, a CFC (chloroflourocarbon), has been replaced with a non-CFC medium such as FM 200.

The operations of ONGC result in the emission of air pollutants and hazardous wastes. With regard to air pollution, the main pollutants emitted include SPM (suspended particulate matter), RSPM (respirable suspended particulate matter), SO_2 (sulphur dioxide), fugitive emissions, NO_x (oxides of nitrogen), and H_2S (hydrogen sulphide).

Figure 1
Group gathering station, Narimanam

The single measure to abate these emissions is the reduction of gas flaring. To reduce emissions of H_2S, this has been coupled with caustic dozing that neutralizes H_2S in oil.

An ETP (effluent treatment plant) has been installed at all establishments. Effluent after treatment is injected in effluent disposal wells at a depth of about 1000 metres. Albeit no water pollutants are discharged, the following parameters in effluents are treated before disposal: pH, oil and grease, and TSS (total suspended solids).

Oily hazardous waste (sludge) is generated during the processing of crude oil and gas at the installation. Authorization for collection and storage of the sludge has been obtained from the TNPCB (Tamil Nadu Pollution Control Board). Presently, 450 MT of sludge is subjected to bioremediation.

Based on a quantitative risk assessment, a disaster management plan was drafted in June 2004, which has since been revised in March 2006. Hazard identification was done while establishing and implementing OHSAS 18001: 1999 at GGS-Narimanam, which was certified in April 2005.

Case study

ONGC commissioned a GGS at Kuthalam village, Nagapattinam district, Tamil Nadu, in August 1989 for crude oil processing, storage, and dispatch facilities. The station was to process crude oil received from Narimanam and other satellite fields, before dispatching it to a refinery of CPCL (Chennai Petroleum Corporation Ltd) at Panangudi, after ensuring that all quality standards were maintained. Subsequently, a gas grid was also established by GAIL India Ltd in the area for supply and distribution. Presently, the GGS is connected to 32 flowing wells and a common ETP.

GGS-Narimanam has facilities to process sweet and sour crude produced from the Narimanam oil field to the tune of 1200 m^3 (cubic metres)/day and 1000 m^3/day, respectively. It has been certified for ISO 9001: 2000, ISO 14001: 2004, and OHSAS 18001: 1999.

Initially, the Narimanam oilfield wells contained sweet crude oil and exhibited the 'self-flow' property. In due course, sour wells were discovered and used for the production of oil. When the wells were on 'self-flow', the quantity of low-pressure gas flared was very small. However, when artificial gas lift operations commenced, a huge quantity of lift gas used for this turned sour, and had to be flared to suit consumer needs. The flaring of sour gas causes air pollution and resource depletion.

Sour gas flaring has been a major concern for a long time. In the absence of consumers for sour gas, various other alternatives had to be tried. These include the following.

- *Gas lift optimization* The quantity of sweet gas used for lifting oil from sour wells has been optimized by following the recommendations of the IOGPT (Institute of Oil and Gas Production Technology). Thus, a marginal reduction of sweet gas turning sour has been achieved, which, in turn, has reduced gas flaring.
- *Sweetening of sour gas* The economics of undertaking this were prohibitive, and hence, the project could not be undertaken on a larger scale.
- *Conversion of gas lift wells to SRP (sucker rod pump) wells* The above proposal was thought of so that sweet gas turning sour in gas lift wells

could be totally eliminated and sweet gas hence produced/saved could be marketed directly. However, the conversion entailed immense capital investment in material selection, procurement of surface and sub-surface units, increased maintenance expenditure, etc.

With the opening up of the economy, the demand for sour gas increased. The initial proposal for selling sour gas was initiated as early as September 2001. As the sales procedures were not crystallized, considerable delays occurred and the final approval for sales was given by the chairman and managing director of ONGC in December 2004. The ONGC management had taken a decision that zero flare/minimum technical flare should be achieved at all production installations, so as to conserve this non-renewable source of energy. The options deliberated were the creation of facility for sweetening of sour gas and sale of sour gas to consumers.

As the first option was deemed not viable, the second option was considered, and accordingly, negotiations for the sale of sour gas with various consumers were made, and in-house modifications of processing equipment was done. As per the contract with the buyer, ONGC had to ensure the following sour gas composition.

- Methane: not less than 75% volume of methane
- Other gaseous hydrocarbons: not to exceed 20% volume
- Non-combustible gases other than HC: not more than 8% volume
- Total sulphur content as H_2S: 4 PPM and 800 PPM
- Moisture: no free water to be present

To ensure this composition, ONGC has made technical modifications at the GGS. An existing separator has been converted as sales gas scrubber and pipeline modifications have been made to route the sour gas to this scrubber. In addition to this, sweet gas, which was being flared, is utilized for the generation of electricity using a gas genset, which has contributed towards savings in the energy bill to the tune of Rs 175 000 per month since February 2006.

In April 2004, the quantity of sour gas being flared was approximately 80 000 SCMD (standard cubic metres per day). As the consumers had to establish infrastructure for processing and utilization, the project commenced in February 2006, and since then, the flaring has been considerably reduced and production has gone up by 77%. Figure 2 shows the amount of sour gas flared, its sales, and utilization at GGS-Narimanam from April 2004 to August 2006.

The following positive impacts have been noted as a result of the project.

- Ambient air quality monitoring reports indicate that the quantity of SO_2 in the atmosphere has decreased from 0.9–1.8 μg (microgram)/m^3 to less than detectable levels.
- After commencement of sales, the amount of flaring has reduced from about 80 000 SCMD to 20 000 SCMD. This reduction has resulted in decreased pollution of the environment.
- Villagers in surrounding hamlets were concerned about the impact of flare on their health, and a reduction in this flaring has allayed their concerns.
- Power generation through sour gas reduces the use of alternative fossil fuels—that is, coal and diesel.

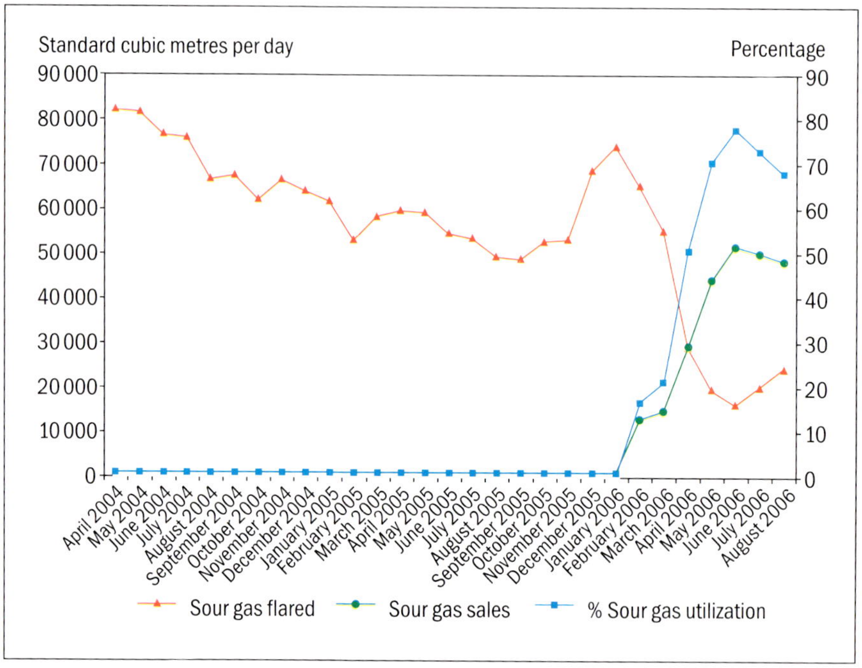

Figure 2
Sour gas flared, sales, and utilized (April 2004–August 2006)

- Huge generation of revenue contributes to the wealth of stakeholders, the state, and central exchequers. This is shown in the following cost–benefit analysis.

Gas quantity sold:	55 000 SCMD
Rate/1000 SCMD:	Rs 2717
Revenue generated/day:	Rs 149 435
Revenue generated/month:	Rs 4 483 050
Revenue generated/year:	Rs 54 543 775

CASE STUDY

at a glance

- ONGC (Oil and Natural Gas Corporation) is one of the largest players in the oil and natural gas industry in the country.
- The company's GSS (group gathering station) is located in Nagapattinam district, Tamil Nadu, and registers an annual production of 72 250 MT (million tonnes) of crude oil and 26.031 MSCM (million standard cubic metres) of natural gas.
- When artificial gas lift operations commenced at GGS-Narimanam, a huge quantity of lift gas used for this turned sour, and had to be flared to suit consumer needs. The flaring of sour gas causes air pollution and resource depletion.
- The ONGC management had taken a decision that zero flare/minimum technical flare should be achieved at all production installations, so as to conserve this non-renewable source of energy.
- The company drew up a contract to sell sour gas and made technical modifications at the GGS to ensure the following composition of the sour gas.
  - Methane: not less than 75% volume
  - Other gaseous hydrocarbons: not more than 20% volume
  - Non-combustible gases other than HC: not more than 8% volume
  - Total sulphur content as H_2S: 4 PPM (parts per million) and 800 PPM
  - Moisture: no free water to be present
- The existing separator has been converted to a sales gas scrubber and pipeline modifications have been made to route the sour gas to this scrubber. Sweet gas, which was being flared, was utilized for the generation of electricity using a gas genset, which has contributed towards savings in the energy bill to the tune of Rs 175 000 per month since February 2006.
- Ambient air quality monitoring reports indicate that the quantity of SO_2 in the atmosphere has decreased from 0.9–1.8 μg (micrograms)/m^3 to less than detectable levels.
- After commencement of sales, the amount of flaring has reduced from about 80 000 to 20 000 SCMD (standard cubic metres per day). This reduction has resulted in decreased pollution of the environment.
- Power generation through sour gas reduces the use of alternative fossil fuels–that is, coal and diesel.

PERFORMANCE
SAFETY
CUSTOMER CARE
ECO-MANAGEMENT
EXCELLENCE EVERYWHERE
At Numaligarh Refinery, we're committed to excel.
In every aspect of our functioning.
It's excellence that's continuous.
Numaligarh Refinery. Synonymous with Excellence
NRL
NUMALIGARH REFINERY LTD.

CASE STUDY 17

Oil and Natural Gas Corporation Ltd, Cauvery Asset

ENVIRONMENT-FRIENDLY ZERO CEMENT DUST EMISSION MODULE

Background

ONGC (Oil and Natural Gas Corporation) is one of the largest players in the oil and natural gas industry in the country. The ONGC Group – comprising ONGC, wholly-owned subsidiary OVL (ONGC Videsh Ltd), and MRLP (Mangalore Refinery and Petrochemicals Ltd) – produced 1.11 million barrels of oil per day in 2005/06. The ONGC Group turnover for the same period was Rs 864.14 billion.

Environmental profile

ONGC has an integrated HSE (health, safety, and environment) policy at the both corporate and facility levels. At the corporate level, 408 installations (more than 99% of field installations and facilities) have established an EMS (environmental management system) based on ISO 14001.

The main programmes and initiatives related to the environment, social responsibility, and sustainable development undertaken by ONGC are quality and HSE certification obtained under ISO 9001, ISO 14001:2004, and OHSAS (occupational health and safety advisory services) 18001; projects under the CDM (clean development mechanism); and the replacement of Halon with a non-CFC (chloroflurocarbon) medium like FM 200. The facility prepared a CDM project for zero flare at GCS (gas-collecting station) Kuthalam, which it submitted to the MoEF (Ministry of Environment and Forests).

Case study

The cementing process and resultant environmental and health hazards

ONGC produces oil and natural gas from wells. Hundreds of wells make up an oil and gas field, and, therefore, there are innumerable cementing operations that need to be preformed in this industry. These wells are drilled from 500–5000 metres, depending on the depth of the presence of hydrocarbon in the earth's layer. The cementing operation is crucial because it needs to isolate the producing layer from the other layers that may hinder its productivity. The role of cementing is vital because it has a direct impact on the health and longevity of any hydrocarbon well. Figure 1 depicts the activities performed during drilling a well and cementing a casing, which is lowered into the drilled hole to protect it from collapsing.

Cementation is a process whereby dry cement is blended with various types of chemicals known as cement additives; this blend is then mixed with water to form slurry that is pumped into the well and is placed between the casing pipe and the well bore. Figure 2 depicts the cementing process.

As can be seen from these diagrams, the cementation process involves a sequence of pneumatic and manual handling of dry cement. The handling of dry cement has scope for various environmental and OHSAS-related issues to arise as the working environment can become uncomfortable and there are chances for process-borne hazards.

This case study primarily focuses on the products used to overcome the present challenges posed by the cementing operations. The nucleus

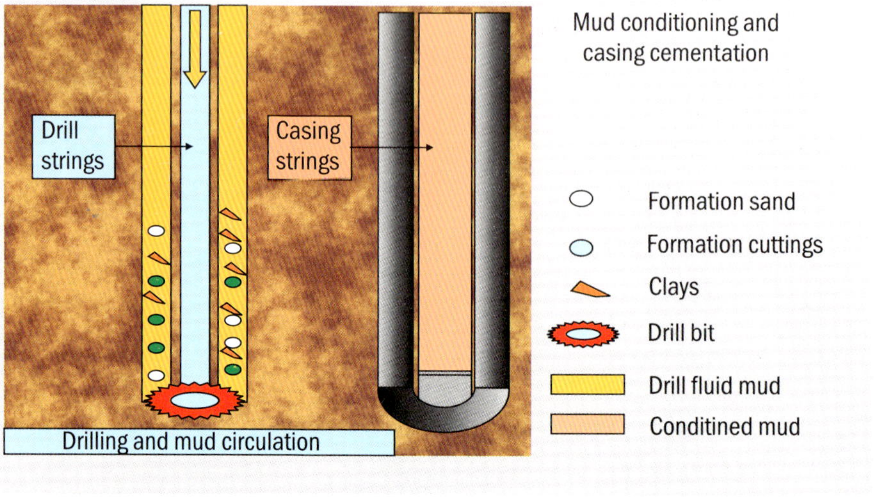

Figure 1
Basic drilling and cementing process

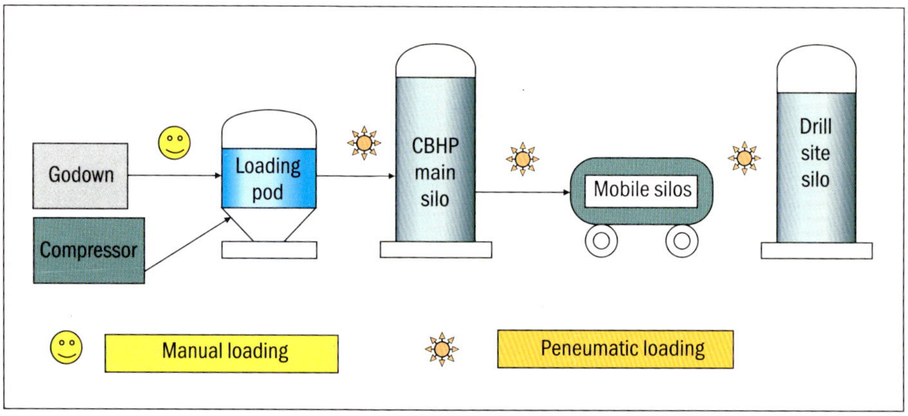

Figure 2
Cementing process

concept of developing this project is to combat environmental pollution, eliminate health hazards, ensure work safety, and bring out exceptional process quality. The shortcomings of the original system were the following.

- Environmental pollution was caused by the blowing of cement dust and cement additive into the air from the surge tank, silo, and open hopper.
- 20–25 kg of cement dust per operation was being blown into the environment.
- Operators and service providers were being exposed to health risks.
- Farms in the vicinity of the well site were being polluted by cement dust.
- Cement quality was being degraded.
- Tonnes of cement was being lost.

ONGC's innovative system

In order to tackle the aforementioned problems, ONGC developed an environment- and user-friendly system by introducing a technically advanced mechanism. This proactive process is divided into two phases

Phase 1

The quality of compressed air in terms of presence of water vapour and other chemicals in vapour form significantly impacts the performance of the system in application, especially when powdered material like cement needs to be transported, as it has a tendency to react with the moisture present in the air. The presence of moisture in compressed air causes adhesion of cement particles to each other. The bigger the particle size, the greater is the pressure requirement for handling/transferring cement; hence, the air volume requirement increases tremendously, which is undesirable as larger volumes of air pose greater difficulties in venting control and carry larger amounts of dust, causing increased dust pollution in the atmosphere.

The first phase of the process, thus, deals with the removal of excess undesirable water content from the compressed air, which is achieved by the inline dehumidifier (Figure 3), developed on the fluid mechanics principle. Figure 4 depicts how the dehumidifier functions.

As shown in Figure 4, compressed air at 4 kg/cm^2 (square centimetres) and 65 °C enters the dehumidifier module, installed between the air compressor unit and air reservoir. The air passes through various cross-sections of the system, circulated by water at ambient temperature. It then experiences a cooling effect (as per the principle of fluid mechanics) and dissipates its heat to circulating water (as per the principle of thermodynamics). The sudden cooling of compressed air condensates

Figure 3
Inline dehumidifier

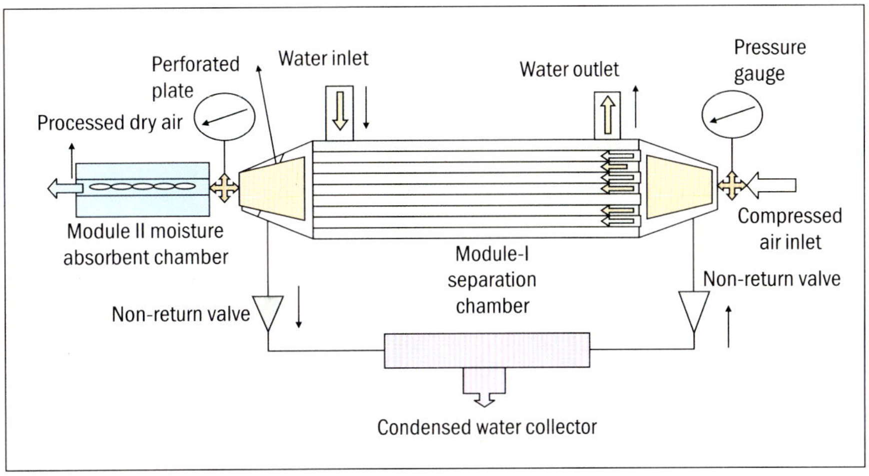

Figure 4
Functionality of inline dehumidifier

moisture in water droplets, which are collected in a chamber and removed from the system. This compressed cooled air is then passed through the secondary absorbent chamber, where the remaining moisture is eliminated by silica gel and charged coal. The dehumidifier effectively eliminates 1500–2000 cm^3 of water condensate from compressed air.

Phase 2

The second phase deals specifically with the cement-mixing process for slurry preparation. In the earlier system, cement and chemical dust in dry powdered form entered the environment in an uncontrolled manner, not only polluting the atmosphere, but also giving rise to health risks. This complete process has been modified and ONGC has developed an automated pollution control module, shown in Figures 5 and 6, which has automated the dry cement supply system for slurry preparation.

As compared with the process depicted in Figure 2, in the automated pollution control module, the manual supply valve is automated and the conventional mixing hopper is replaced by a concealed regulated hopper. Dry cement supply is also regulated by automated sensing of the level in the surge tank, which ensures adequate quantity and maintains the air vent for restricting dust flow. The dust trap automizer unit collects cement dust from air dust vent centrifugally, and the remaining particles are adsorbed and diffused with atomized water particles. An average of 5–10 MT (million tonnes) of cement dust was trapped per annum.

Benefits

The main benefits of this case study are that environmental pollution is reduced because of zero dust emission, and occupational health risks are completely eradicated. The innovations described above controlled environmental pollution by reducing the air volume requirement for the transfer of cement and addressed the problems of uncontrolled flow of cement dust at the mount of the mixing hopper, pollution from the surge tank and silo, and lack of control and monitoring in supply of cement to the mixing hopper. Addressing these problems also eliminated health hazards for employees. Further, the innovations resolved the auto-retardation tendency of cement when it is exposed to high humidity and

Figure 5
Fully automated pollution control module

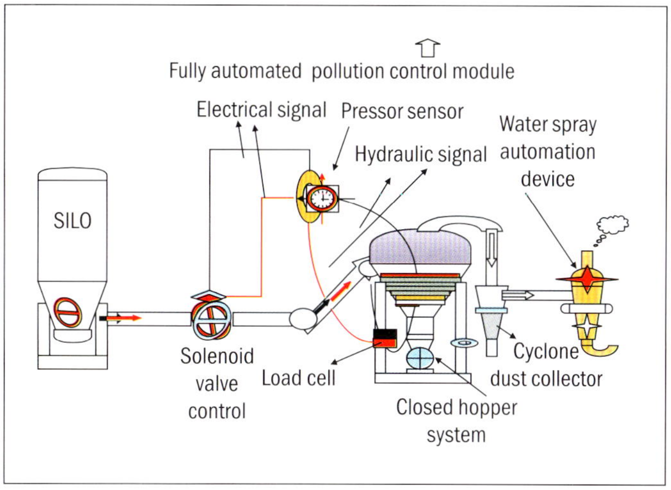

Figure 6
Process diagram of fully automated pollution control module

moisture, and the subsequent degradation of cement quality during handling and storage pneumatically in dry powdered form and rectified the lack of control in maintaining cement level in the mixing module hopper. Thus, it also provides improved quality cementing operations and much less process-related waste.

Cost-benefit analysis

As the material used in this project is part of the system already in place, the implementation cost was only of the modifications and fabrications that took place. Therefore, the total project cost was Rs 200 000. However, the cost of a new system would cost approximately Rs 600 000.

This cost of this project would be recovered in few quality cementation jobs, which reduces the cement repair job cost and the project will earn profit both tangible and intangible. Since the implementation of the complete project, the company has saved approximately Rs 2 million.

CASE STUDY

at a glance

- ONGC is one of India's largest oil and natural gas producing companies with an annual turnover of Rs 10.82 billion for 2005/06.
- The case study focuses on installation of a dehumidifier to reduce the water content in the compressed air and the development of an automated pollution control module for slurry preparation during the cementing operation.
- These innovations address the following issues
  - Environment pollution by blowing cement and cement additive into the surrounding air.
  - 20–25 kg of cement dust per operation blown into the environment.
  - Cementing operations exposing operators and service providers to health risk.
  - Presence of moisture in compressed air, which is a media for pneumatic transfer, causes adhesion of cement particles to each other, creating more pressure in the air for handling.
- In the first phase of the two-phase process developed by ONGC to address the above-mentioned problems, a dehumidifier was installed so that moisture is removed in the primary chamber and the remaining moisture is removed by silica gel and charge coal absorbent chambers.
- In the second phase, an automated supply valve, concealed regulated hopper, automated level sensing in surge tank, and dust trap-atomizer unit were introduced.
- The two-phase process led to the elimination of 1500–2000 cm^3 (cubic centimetres) of water condensate from the compressed air, the trapping of an average of 5–10 MT (million tonnes) of cement dust per annum, and pollution control by zero-cement-dust emission.

CASE STUDY

18

Rashtriya Ispat Nigam Ltd

COMBATING GLOBAL WARMING

Background

Rashtriya Ispat Nigam Ltd's VSP (Visakhapatnam Steel Plant) is located in Visakhapatnam, Andhra Pradesh. The company had an annual turnover of Rs 88.72 billion for 2005/06.

Environmental profile

VSP has an environment policy and department at both the corporate and facility levels. It is the only steel plant in India to be certified with all the international standards—ISO 9001, 14001, and 18001. The company has received several awards in the past including the National Energy Conservation Award from the Ministry of Power, Government of India for three consecutive years, the Leadership and Excellence Award for SHE (safety, health, and environment) Management in 2005 by the CII (Confederation of Indian Industries), and the Golden Peacock Award for Environmental Management by the World Environment Foundation in 2006.

The facility has a wide array of air pollution control systems. Systems to control SPM (suspended particulate matter) include 27 cyclones, 80 scrubbers, 15 ESPs (electrostatic precipitators), 58 bag filters that are provided wherever raw materials are transferred through belt conveyors, and six gas cleaning and recovery systems. Regular monitoring of the ambient air is done to ensure compliance with norms. Low sulphur boiler coal is used to abate SO_2 (sulphur dioxide) emissions. Over the past three years, VSP has reduced about 13 000 tonnes of SO_2. Noise is controlled from the design board itself—all rotating equipment are designed such that they do not exceed 90 dBA (decibel [acoustic]) of noise. For high-speed compressors, turbo-generators, and turbo-blowers, acoustic enclosures are provided. Employees are also supplied with ear-muffs and ear-plugs.

Toxic effluents are generated only in the coke ovens and coal chemical plant, and these are treated first in the mechanical, biological, and chemical treatment plant, and then in the ETP (effluent treatment plant), after which they are discharged into the sea. All parameters are kept within the norms. There is an STP (sewage treatment plant) wherein effluents are treated and recycled through reverse osmosis process for cooling applications in the rolling mills, steel melt shop, etc. Further, ash slurry is sent to the ash pond and water is recycled. One of the four boilers has been equipped with a dry fly ash handling, storage, and delivery system with a capacity of 3.42 tonnes per hour to enable the manufacture of fly-ash-based products such as fly ash bricks. This initiative has also reduced the ash slurry load on the ash pond and will soon be replicated at the other boilers.

Hazardous items such as activated sludge, tar sludge, benzol muck, and acid sludge are charged in coke oven batteries and incinerated. These hazardous wastes get converted to pure carbon, which is equivalent to the coke produced from the imported coking coal. Waste oil and non-ferrous waste are sold to parties authorized by the MoEF (Ministry of Environment and Forests). A unique system was developed for the mechanized mixing of hazardous benzol sludge with dry coke dust in a concrete mixer to eliminate manual mixing (shown in Figures 1a and 1b). The semi-dry mixture is packed in polythene bags and then loaded on a conveyor for charging into coke oven batteries as seen in Figure 1b.

Resource conservation activities are also undertaken at the facility. Major water conservation projects implemented include the recycling of secondary spray cooling water, recycling of sewage water from the STP, and the elimination of drift losses from the cooling towers. Specific water consumption per tonne of liquid steel has thus reduced from 3.96 m^3/TSL (tonne of liquid steel) to 2.56 m^3/TLS. Moreover, 122 rainwater harvesting projects have been carried out during 2002–06.

A few noteworthy energy-saving projects include the reduction of heat consumption in coke oven batteries by optimization of moisture in coal blend and cross-wall temperature, heat consumption reduction in the TPP (thermal power plant) by replacing damaged baskets of air pre-heaters in boilers, and heat consumption reduction in BF (blast furnace) by optimizing the combustion in stoves.

VSP organizes several activities to promote and spread environmental awareness. About 330 employees were trained in 12 programmes on waste

1a

1b

Figures 1a and 1b
Mixing of benzol sludge with dry coke dust before and after installation of mechanized concrete mixer

management and 115 on programmes on operation and maintenance of pollution control equipment. Competitions are organized for employees, housewives, and school children. Green awards are distributed to 50 employees every year, recognizing their initiatives towards improving the environment through water conservation, waste recycling, afforestation, energy conservation, etc. VSP has held seminars such as 'Utilization of fly ash and Green cities—plan for a planet' to spread awareness and generate discussion on the environment.

Several initiatives have been taken by VSP in terms of CSR (corporate social responsibility), including promotion of literacy, girl-child education, special training and rehabilitation for the mentally and physically disabled, support for tribal students, pilot projects on the prevention and control of HIV/AIDS, carrying out of extensive medical and health camps, provision of infrastructure in the surrounding villages, provision of support during natural calamities, and efforts towards the preservation and encouragement of tribal culture. The installation of biogas plants and solar water heaters in three peripheral villages is a specific example of CSR projects undertaken by VSP. The company allocates 0.5% of its profits towards these activities.

Three projects are being implemented under the CDM (clean development mechanism). The first project entails the use of pulverized coal injection in the blast furnace and reduces an estimated 175 000 tonnes of CO_2 (carbon dioxide) per year. The second is the waste heat recovery project in the sinter straight line cooler that aims to reduce about 100 000 tonnes of CO_2 per year. Finally, the blast furnace waste heat recovery system project is expected to lower CO_2 emission by 130 000 tonnes per year.

Case study

Global warming and climate change are the issues that have found themselves at the forefront of environmental discussion and are painting a grim picture for the global environment scenario. Of all the GHGs (greenhouse gases) that lead to global warming, CO_2 has been identified as the largest contributor (55%). The VSP has thus taken several innovative and constructive steps to identify and minimize its contribution of GHGs, especially CO_2, to the atmosphere.

In order to drastically cut down its GHG emissions and help combat global warming, VSP identified three specific areas in which to take action: (1) reduction in the emission of CO_2, (2) reduction in emission of SO_2, (3) reduction in consumption of ODS (ozone-depleting substances) and CFCs (chloroflurocarbons).

Reduction in CO_2 emission

In a step towards reducing CO_2 (per tonne of liquid steel), VSP laid emphasis on reducing the coke rate in blast furnaces. Hence, the gradual and successful reduction in the coke rate has resulted in an average reduction of CO_2 emissions of approximately 220 000 tonnes/year. Similarly, efforts have been made to reduce CO_2 emission from coke oven gas consumption by increasing the efficiency of the burners. Total CO_2 was also reduced by presenting ingress of air into the coke oven gas network from the coke oven battery onwards. As a result, CO_2 emission was reduced by 68 tonnes/year.

The conservation of energy and other resources is a highly effective indirect method towards reducing CO_2 emission. The generation of energy through burning of fuel gases or coal emits CO_2. Similarly, the recycling of raw materials such as coal, iron ore, and limestone is encouraged on the site, not only because it conserves natural resources, but also because it leads to savings at numerous stages of the process such as mining, transportation, crushing/processing, etc. The recycling of intermediate products like coke breeze and nut coke lowers the energy requirement in producing that material. Based on data from 2002/03 to 2005/06, the reduction of electrical energy consumption has led to an average annual reduction of 14 950 tonnes/year of CO_2 equivalent. Similarly, approximately 445 tonnes/year of CO_2 equivalent has been reduced due to water conservation initiatives, 1540 tonnes/year through reduction in consumption of lubrication oil, 364 tonnes/year through reduction in specific refractory consumption, and an average of 68 tonnes/year as a result of reducing coke oven gas consumption.

Afforestation is another way by which CO_2 emissions can be slashed. Not only do trees provide a cooling effect on the environment, but it has also been estimated that a single tree can sequester approximately 11.8 kg of CO_2 each year. VSP planted 85 000 trees in 2005/06 and a total of 3 819 000 trees since 1996. Calculations based on data from 1996/97 – taking five years since plantation of the saplings as the average age to attain maturity and a CO_2 sequestration potential of 11.8 kg/tree – reveal that the average annual reduction of CO_2 equivalent amounts to approximately 3345 tonnes/year.

Reduction in SO_2 emission

The main source of emission of SO_2 is the sulphur content in the boiler coal that is used for power generation in VSP's thermal power plant. Hence, it was decided that low-sulphur coal be used in all the boilers—the percentage of sulphur in coal has reduced from 0.45% in 2002/03 to 0.31% in 2005/06, despite the lower cost of coal containing higher amounts of sulphur. The use of superior boiler coal has led to an average annual reduction of SO_2 emissions of 4368 tonnes/year.

Reduction in consumption of ODS and CFCs

The consumption of ODS has been reduced by 9.74% (that is, 414 kg from 4254.75 kg in 2004/05 to 3840.24 kg in 2005/06. Moreover, 32 of 141 Halon-1211 extinguishers have been taken out of service in 2005/06, resulting in the removal of 160 kg of Halon-1211 and a CO_2 equivalent of 297.6 tonnes. Additionally, 45 water coolers and 23 panel air conditioners working on ODS (R-12 refrigerant) have been replaced with a non-ODS refrigerant (R-13a and R-22) at a cost of Rs 1.26 million. This has, however, led to a saving of 34 kg of R-12 in 2005/06, which translates into a CO_2 reduction of approximately 364.5 tonnes.

Cost benefit

The initiatives and techniques employed by VSP to reduce CO_2 emission to combat global warming prove to be highly rewarding. Not only has the above-mentioned project led to an overall reduction of about 223 591 tonnes/year of CO_2 equivalent, but has also brought monetary benefits. Financial investment required to fund the project amounted to Rs 98.46 million, while returns and savings amount to Rs 408 million.

CASE STUDY

at a glance

- Rashtriya Ispat Nigam Ltd's VSP (Visakhapatnam Steel Plant) in Andhra Pradesh had an annual turnover of Rs 88.72 billion for 2005/06.
- VSP has taken several steps towards combating the growing threat of global warming and GHG (greenhouse gas) emission.
- Measures included, primarily, reducing CO_2 (carbon dioxide) and SO_2 (sulphur dioxide) emission, and cutting the consumption of ODS (ozone-depleting substances) and CFCs (chloroflourocarbons).
- To lower its CO_2 emissions, VSP has enhanced process-efficiency and reduced the coke rate in blast furnaces, implemented energy and resource conservation techniques (including reduction in electrical energy consumption, water consumption, lubrication oil consumption, and refractory consumption), and afforestation.
- SO_2 emission has been reduced by using superior boiler coal that has a lower sulphur content percentage.
- The reduction in consumption of ODS, removal of Halon-1211 extinguishers, and replacement of water coolers and air conditioners working on ODS refrigerants with those that work on non-ODS refrigerants have also led to a reduction of CO_2 equivalent.
- These efforts have collectively resulted in the overall reduction of about 223 591 tonnes/year of CO_2 equivalent and sizeable monetary savings of Rs 408 million, as against an investment of Rs 98.46 million.

CASE STUDY

19

Sona Koyo Steering Systems Ltd

Elimination of hazardous waste generation by process modification

Background

Sona Koyo Systems Ltd, hereafter referred to as Sona Koyo, is the forerunner of the Sona Group of Companies. It is currently the largest manufacturer of steering systems for the passenger car and utility vehicle market in India. Annual production at their facility in Gurgaon, Haryana, was 730 000 steerings for 2005/06 and the annual turnover for the same period was over Rs 4000 million.

Environmental profile

Sona Koyo has taken several initiatives related to the environment, social responsibility, and sustainable development. In order to augment the socio-economic development of the communities around its facility, Sona Koyo has taken up several community development programmes that focus on child care, health care, education, and women's empowerment. It has opened a dispensary that treats about 50 patients free of cost daily; organizes vaccination drives; continually upgrades facilities at schools and strives to improve the quality of education; etc.

Some such steps taken at their facility include providing a drainage system in the scrap yard to collect oil and coolant; metal cladding the scrap yard to prevent soil and water contamination; installing an incinerator for destroying wastes; setting up an ETP (effluent treatment plant), STP (sewage treatment plant), and hazardous waste collection pit with secured storage; and partaking in projects (with TERI, New Delhi) towards rainwater harvesting.

All air pollutants emitted by Sona Koyo's facility at Gurgaon, Haryana are frequently checked and kept within the permissible limits of the CPCB (Central Pollution Control Board) norms, as shown in Table 1.

Having established its state-of-the-art water treatment facilities, Sona Koyo is a zero-discharge plant. Treated waste water is used for gardening, and feasibility studies are under way for using treated ETP water in cooling towers.

Hazardous wastes generated at the facility and their disposal methods are listed in Table 2.

TABLE 1
AIR POLLUTANTS AND POLLUTION CONTROL BOARD LIMITS

List of air pollutants	*Level in the premises*	*CPCB limits*	*Measures taken to abate them*
(SO_2) Sulphur dioxide	8.65	120	Emission is within the HSPCB limits
(NO_x) Oxides of nitrogen	21.5	120	Emission is within the HSPCB limits
(SPM) Suspended particulate matter	26.5	500	Emission is within the HSPCB limits
CO (Carbon monoxide)	1132.5	5000	Emission is within the HSPCB limits

CPCB – Central Pollution Control Board; HSPCB – Haryana State Pollution Control Board

TABLE 2
HAZARDOUS WASTE DISPOSAL

Type of waste	*Method of disposal*
Waste from paints, thinner, varnish	Incineration
Oily cotton waste, hand gloves	Incineration
Used hydraulic oil	Sold to authorized dealer
Used quenching oil	Sold to authorized dealer
Hazardous waste sludge	Hazardous waste collection pit and secured landfill
Used battery (acid/alkaline/slurry)	Authorized dealer/supplier takes back the old batteries during the time of supply of new batteries
Empty paint container	Cleaned in ETP as per operational control procedure and sold outside
Empty oil drums	Cleaned in ETP as per operational control procedure and sold outside
Empty chemical containers	Cleaned in ETP as per operational control procedure and sold outside
Ash from the incinerator	Stored temporarily in a safe place, to be used in construction as per operational control procedures

ETP – effluent treatment plant

Along with increasing plant capacity, focus is laid on maximizing equipment productivity and resource efficiency. Periodic audits are carried out for maximizing plant productivity and reducing specific consumption of key resources like energy, water, and chemicals.

Case study

Safety, health, and environment at Sona Koyo

Sona Koyo has an SHE (safety, health, and environment) department that conducts daily safety patrol rounds to eliminate unsafe conditions and actions in the different processes such as machining, grinding, heat treatment, turning, boring, assembling, testing, and painting. One such issue raised after a safety patrol round in the paint shop was to eliminate the usage of chromic acid in the painting process.

Hazards of chromium usage

Chromium is a well-known toxic substance. Wastes containing chromium compounds are classified as 'hazardous' under the Hazardous Waste Management Rules, 1989.

The usage of chromium results in water and soil pollution. It enters the human body through inhalation of chromium mist, eatables, polluted drinking water, or contact with skin during handling. It also has the capacity to alter genetic material and can cause cancer.

Recognizing the potential health hazards of this substance, Sona Koyo has taken initiatives to eliminate chromium-based passivation, a step in the painting process.

Moreover, the effluent from the paint shop revealed the presence of a high amount of chromium, which caused problems in the efficiency of the ETP. The excess of hexavalent chromium present in the paint shop effluent is treated with $NaHSO_3$ (sodium bisulphate) in the ETP and converted to trivalent chromium, which forms a green coloured hazardous sludge. Thus, it was decided that in order to eliminate chromium in the effluent, chromium consumption must be eliminated altogether in the paint shop.

Conventional painting process

Painting is done to prevent corrosion and give an aesthetically pleasing finish to products. Figure 1 depicts the process flow chart of the paint shop process.

Degreasing (Step 2 in Figure 1) is a physical process wherein caustic cleaner is used by power spraying on the product to remove excess grease and oil on the product. Surface activation (Step 4) is a conditioning rinse process for the steel surface. TiO_2 (titanium dioxide) is used as a conditioning agent. This is an oxidation process that converts the ferrous iron present on the steel to ferric oxide. Phosphating (Step 5) is done to remove excess free iron on the steel surface. The iron is converted to insoluble form and gets deposited in sludge. Passivation (Step 7) is the final step in the chemical treatment of the paint shop and requires that the surface of the steel should be completely free of ferrous iron.

Passivation and the use of chromic acid

The passivation step of the painting process entails removing exogenous iron or iron compounds ('free iron') from the surface of steel by chemical dissolution without significantly affecting the steel. An acid solution is usually used to remove surface contamination. Chromic acid (about two litres per day) is typically used in passivation.

The basic nature of chromic acid is its oxidizing nature—in the passivation process, it oxidizes the free iron to iron oxide, and thus prevents corrosion. Chromic acid also enhances the spreading of the paint in the material and acts as an extra shell between the zinc phosphate coating and the paint.

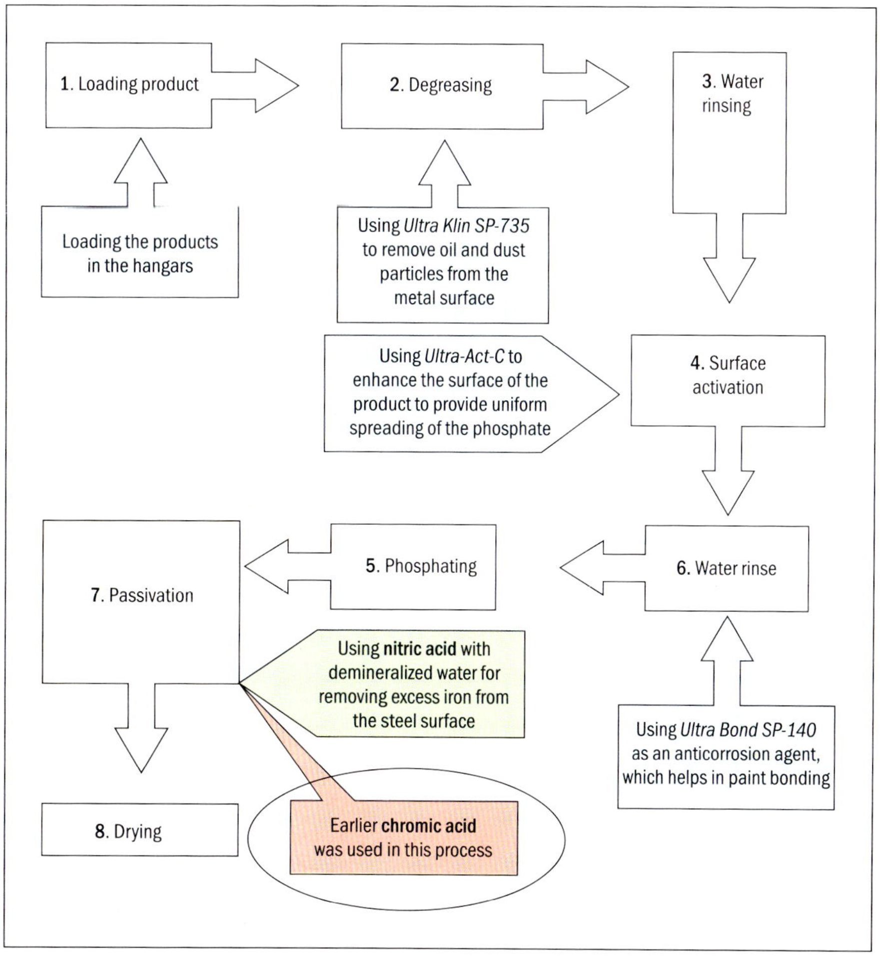

Figure 1
Paint shop process

The steel product is thus usually treated with hexavalent chromium, which oxidizes ferrous iron to ferric oxide as shown below. Due to the loss of three electrons, the hexavalent chromium gets reduced to trivalent chromium and deposits at the bottom of the tank.

Fe	+	Cr^{6+}	→	Fe_2O_3	+	Cr^{3+}
(ferrous iron)		(hexavalent chromium		(ferric oxide)		(trivalent chromium)

As the use of hexavalent chromium has been banned in India, all other automobile and auto component industries have been using trivalent chromium for passivation as shown below. The trivalent chromium not only causes treatment problems in the ETP, but also a larger amount of it is required as compared to hexavalent chromium.

Fe	+	Cr^{3+}	Fe_2O_3	+	Cr^{3+}
(ferrous iron)		(trivalent chromium)	(ferric oxide		(trivalent chromium)

Modification of the painting process

After conducting intensive research and a series of tests, Sona Koyo designed an efficient modified painting system in association with the paint shop. Each step in the painting process was modified, as detailed in Table 3, and followed by over 100 trial runs during a six-month period.

As shown in Table 3, Sona Koyo used nitric acid in the passivation process as it fulfilled the requirements of eliminating the free ferrous iron on the surface of steel products. As shown below, when free ferrous iron reacts with nitric acid, it forms ferric oxide and water. Thus, the excess free iron from the surface of the steel product is remove d in the form of ferric oxide:

Fe	+	HNO_3	Fe_2O_3	+	H_2O
(ferrous iron)		(nitric acid)	(ferric oxide)		

After testing, it was found that 100 mililitres of nitric acid per 900 litres of DM (demineralized) water is sufficient for passivation requirements for a day. Further, the painted products met the quality requirement and the efficiency of the product retained is equivalent to chromic acid usage—the modified method passed all the necessary tests such as the adhesion test, paint film thickness check, and salt spray test, and no problems such as cissing, paint flow, orange peel, and blistering were reported.

Benefits of the modified system

Substituting the use of chromic acid in passivation with nitric acid not only met all the quality requirements, but also presented numerous benefits. By eliminating chromic acid, groundwater pollution due to chromium leaching is prevented. Effluents are no longer contaminated with chromium, and treated ETP water can be reused in cooling towers. Chemical consumption in the ETP for chromium treatment is eliminated, which, in turn, eliminates the hazardous green coloured sludge from forming in the ETP. Moreover, it led to safe working conditions and prevention of occupational health hazards.

TABLE 3
TRIAL-BASED MODIFICATIONS TO THE PAINTING PROCESS

Steps in the painting process	*Before modification*	*Trial - 1*	*Trial - 2*	*After modification (optimized process)*
Degreasing	1 kg of degreasing agent in 100 litres of soft water. Temperature 15 °C–25 °C	2 kg of degreasing agent in 100 litres of soft water. Temperature 65 °C–85 °C	4 kg of degreasing agent in 100 litres of soft water. Temperature 45 °C–55 °C	2 kg of degreasing agent in 100 litres of soft water. Temperature 45 °C -55 °C
Surface activation	0.5 kg of surface activator agent per 1000 litres soft water. Temperature 25 °C–35 °C pH 8-10	1.2 kg of surface activator agent per 1000 litres soft water. Room temperature 15 °C -25 °C; pH 7–9	2 kg of surface activator agent per 1000 litres soft water Room temperature. pH 6–7	1.2 kg of surface activator agent per 1000 litres soft water. Room temperature. pH 7–9
Phosphating	Zinc phosphate 9 litres per 900 litres of soft water. Temperature 45 °C–55 °C	Ultra-bond SP 18 litres per 900 litres of water. Temperature 85 °C–95 °C	Ultra-bond SP 18 litres per 900 litres of water. Temperature 85 °C– 95 °C	Zinc Phosphate 18 litres + Sodium nitrate 0.5 litres per 900 litres of soft water. Temperature 85 °C–95 °C
Passivation	Hexavalent chromium 2 litres per 900 litres of water. pH 5–6. Room temperature	2 litres of Trivalent chromium/$KMNO_4$ per 900 litres DM water. Room temperature	Amine alkali phosphate 4 kg per 900 litres of water. Room temperature	100 ml nitric acid 900 litres of DM pH 4-6. Room temperature

DM - deminearlized; $KMNO_4$ - potassium permanganate

The use of nitric acid has also brought about cost benefits. While about 2 litres of chromic acid was used per day in the earlier process, only 100 mililitres of nitric acid are used in the modified process. Sona Koyo spends only Rs 272 per annum on nitric acid, as compared to Rs 102 000 on chromic acid consumption (in addition to Rs 25 500 on sodium-meta bisulphate), resulting in a cost saving of Rs 127 228 per year ([Rs 102 000 + Rs 25 500] – Rs 272 = Rs 127 228).

CASE STUDY

at a glance

- Sona Koyo Systems Ltd is the largest manufacturer of steering systems for the passenger car and utility vehicle market in India, with an annual turnover of over Rs 4000 million for 2005/06.
- The company's SHE (safety, health, and environment) department raised the need of eliminating the use of chromic acid in the painting process, due to its hazardous nature, leading, not only to environmental pollution, but also putting the health and safety of employees at risk.
- A modified painting process was suggested after substantial R&D (research and development) efforts, wherein the key feature was the substitution of chromic acid for nitric acid in the passivation step of the painting process.
- The elimination of chromic acid led to the prevention of groundwater pollution, reuse of ETP water in cooling towers, creation of a safe and healthy working environment for employees and stopped contamination of effluents with chromium.
- The use of nitric acid in place of chromic acid brought about a cost saving of Rs 127 228 per annum.

CASE STUDY

20

Toyota Kirloskar Motors Pvt. Ltd

REDUCTION OF AIR EMISSIONS TO MINIMIZE THE IMPACT ON THE ATMOSPHERE

Background

TKM (Toyota Kirloskar Motors) Pvt. Ltd, a subsidiary of the Toyota Motor Corporation, Japan, is a manufacturer of multi-purpose vehicles and cars. Annual production at their Bidadi facility in Bangalore in 2005 was 60 000 vehicles (increased from 45 000 the year before). Since its establishment in December 1999, TKM has sold approximately 185 000 units. TKM's turnover in 2005 was over Rs 40 million.

Environmental profile

TKM, as a part of its environmental policy, is committed to promoting environmental awareness and knowledge among its team members through continual education and job-specific training. It ensures compliance with legal and other requirements to which the company subscribes; annually reviews its environmental objectives and targets to ensure better environmental performance; and has established programmes that strive to conserve energy, natural resources, and flora and fauna.

To tackle air pollution – in the form of SPM (suspended particulate matter) and noise – TKM has provided filters and ample chimney height as well as enclosures in areas where high levels of noise is generated. The main water pollutants discharged are BOD (biochemical oxygen demand), pH, COD (chemical oxygen demand), TSS (total suspended solids), and TDS (total dissolved solids). Waste water is treated through an on-site waste water treatment plant, which ensures that the major parameters are well within internal limits that are 20% more stringent than the legally defined limits. Details of the generation and disposal of hazardous waste are outlined in Table 1. Paper, plastic, and metal generated by the facility are recycled, while glass, dust, and cotton are disposed off at the municipal landfill site, and wood is reused.

TKM complies with the rules set under the Water Act, Air Act, and Hazardous Waste Rules. Environmental risk management and disaster management plans are in place as well.

Case study

Primary air emissions at TKM

In order to achieve its environmental goals of decreased pollution to the environment, TKM established a five-year environment action plan. One of the key elements of this plan is to improve atmospheric air quality by reducing air emissions. The major sources of air emissions are VOCs (volatile organic compounds); CO_2 (carbon dioxide) due to consumption of electricity and LPG (liquefied petroleum gas) and also due to consumption of fuel for logistics activities.

VOC emission reduction

VOCs are organic compounds that evaporate easily. When released into the atmosphere, VOCs contribute to smog formation and have a direct adverse effect on human health. They are liberated into the atmosphere by the usage of paints and solvents during the painting process of vehicles.

TABLE 1
HAZARDOUS WASTE GENERATION AND DISPOSAL

Type of hazardous waste	Source	Disposal method
Paint sludge	Paint shop	Incineration by HAAT Incinerators Indian Pvt Ltd and ash stored at TKM
Phosphate sludge	Paint shop	Storage at hazardous waste storage area of TKM
Chemical sludge	WWTP	Storage at hazardous waste storage area of TKM
Oil-contaminated waste	Company wide	Disposed through authorized contractors
Empty containers	Company wide	Disposed through authorized contractors
Waste oil	Maintenance	Disposed through authorized recyclers
Spent solvents	Paint shop	Recycling through Komathi Industries
Distillation residue	Processor	Incineration by HAAT incinerators
Incinerated ash	Processor	Storage at hazardous waste storage area of TKM
HFO sludge	Engineering department	Disposed through authorized recyclers
Electronic waste	Company Wide	Disposed through authorized contractors

HFO – heavy fuel oil; WWTP – waste water treatment plant; TKM – Toyota Kirloskar Motors

In order to reduce emission of VOCs, TKM drew up an activity plan consisting of three stages: robot flushing, manual cleaning, and recovery improvement.

Robot flushing

This method entails the use of purge thinner to flush paint colour, then the use of shaft thinner to clean the gear pump, followed by the use of cup thinner to clean the bell cup from inside, and, finally, the use of back face thinner to clean the bell cup from outside. Robot flushing was done after the spraying of each body. This activity reduced VOC emissions from 48.4 g/m^2 (gram per square metre) to 45.6 g/m^2.

Manual cleaning

TKM changed their cleaning method from spraying solvent with a spray gun to the wiping method, which led to the reduction of VOCs to 42.94 g/m^2.

Recovery improvement

In the recovery system, solvent is used in the booth for activities like calibration and cleaning. TKM found various ways of achieving recovery improvement such as collecting paint used during calibration and improving dirty thinner recovery from the robot hopper. Earlier, all viscosity-measuring tools were cleaned in one container and dirty thinner was sent to the recycle unit. After implementing Kaizen activities in the process, each colour tool is dipped in one individual colour container and 100% solvent is reused . Thus, VOC emission reduced to 41.29 g/m^2.

Reducing CO_2 emissions by reducing electricity and LPG consumption

The generation of electricity through thermal power plants and independent DG sets lead to the emission of CO_2. At TKM, about 42% of electricity is purchased from BESCOM (Bangalore Electricity Supply Company Ltd) and 58% of electricity is generated using DG sets.

TKM has an energy management team that sets and reviews targets monthly. To reduce LPG consumption, TKM has optimized electricity consumption, installed a variable frequency drive for high torque machines, optimized compressor-operation-based plant, compressed air requirements, provided optimizers in compressed air lines, replaced corrugated sheet roofing with transparent acrylic sheets for eliminating lighting requirement in the plant during the day time, installed solar water heaters at the canteen and dormitory, etc. The emission reduction as a result of these initiatives is shown in Figure 1.

Reducing CO_2 emissions through logistics

External logistics

External logistics are those activities in which trucks are used for the movement of parts from the suppliers to TKM. TKM operates 119 trucks from 69 suppliers. To minimize CO_2 emissions caused by trucks, TKM sources most of its parts (about 71%) from suppliers near its plant (Figure 2). To minimize the impact on the environment due to CO_2 emissions because of logistics from plants outside the region of its operation, TKM carefully studied the route taken by the trucks to minimize distance travelled, optimized load efficiency, assessed the skills

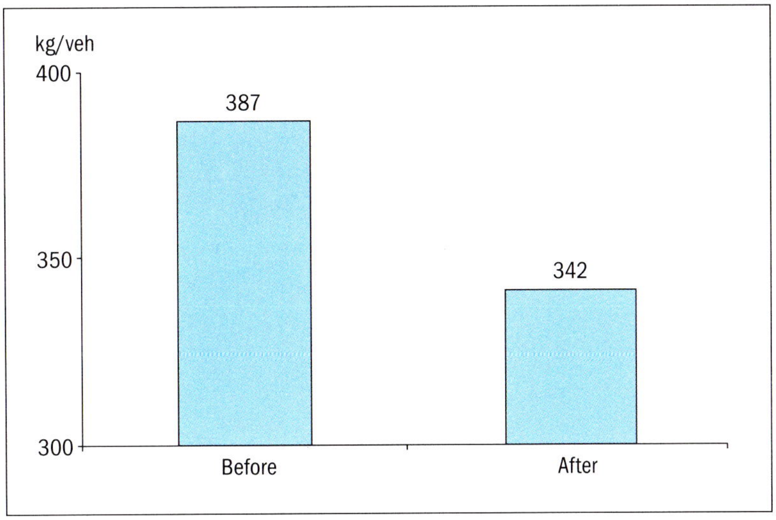

Figure 1
CO_2 emission (electricity and LPG consumption)

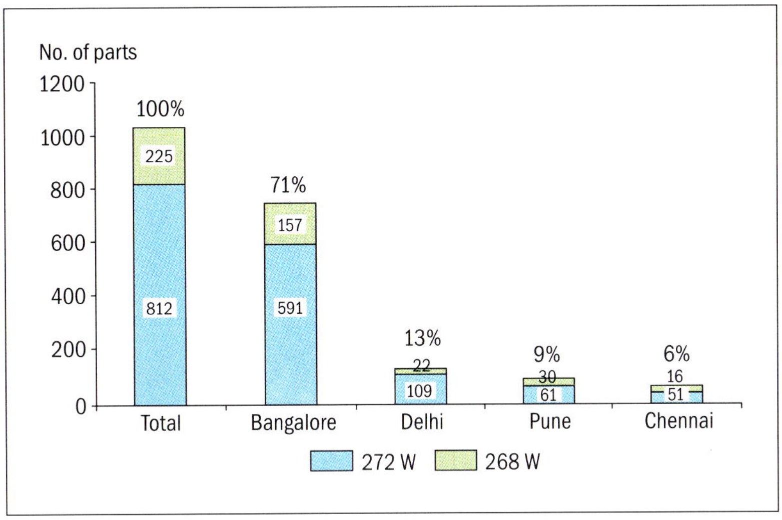

Figure 2
CO_2 emission reduction from external logistics

of the drivers, used the minimum possible number of trucks, and ensured fuel quality. As a result, total kilometres covered per year were reduced by 653 671, and total fuel consumption per year was reduced by 134 316 litres, leading to a total CO_2 emission reduction of 448 tonnes per year.

Internal logistics

Internal logistic activities are those in which diesel-run forklifts are used for the movement of materials within the TKM premises. To minimize CO_2 emitted as a result of this, TKM studied the routes of forklift movement and minimized unnecessary movements, prepared a timetable for supplying materials so that optimal containers/materials are carried per forklift, allocated forklifts based on efficiency so that those with higher efficiency are allocated to processes with greater movement, etc. As a result, CO_2 emissions reduced from 51 tonnes per month to 35.7 tonnes per month.

Benefits of the programme

The initiatives described have resulted in decreased pollution of the environment as they reduced the emission of GHGs (greenhouse gases) into the atmosphere. The case study advocates the reduction in air emissions and conservation of natural resources. Moreover, the aforementioned activities also have a direct impact on employees and TKM as a whole, as the company was able to minimize the cost of production of vehicles, and in turn, increase profits and provide better salary packages and enhanced benefits to employees. Reducing emissions caused by internal logistics also led to reduced air emissions inside shops, which resulted in improved atmospheric conditions. All these measures have a positive impact on the health of employees.

CASE STUDY at a glance

- TKM (Toyota Kirloskar Motors) Pvt. Ltd is a subsidiary of the Toyota Motor Corporation, Japan.
- TKM is a manufacturer of multi-purpose vehicles and cars, with a production facility in Bidadi, Bangalore.
- The case study focuses on reducing the effect on the environment by its manufacturing facility through air emission reduction.
- Major sources of air emissions are VOCs (volatile organic compounds) and CO_2 (carbon dioxide) due to consumption of electricity and LPG (liquefied petroleum gas) and also due to consumption of fuel for logistic activities.
- VOC emissions are reduced as the consumption of thinner by each line is monitored, and consumption is visualized.
- Air emission through logistics and fuel consumption is minimized through meticulously recording of fuel issued and minimizing movement of trucks and forklifts.

CASE STUDY 21

Vishay Components India Pvt. Ltd

Greening electronic components

Background

VCIPL (Vishay Components India Pvt. Ltd) is owned by Vishay Intertechnology Inc., a US-based multinational and one of the world's largest manufacturers of discrete semiconductors and passive components. It is the largest manufacturer of passive components in the country, covering various types of resistors and capacitors that are used as the building blocks for all electronic circuits. It generates revenue of Rs 1200 million, and more than half the production is exported all over Asia, Europe, and the US through VCIPL's global sales and logistics network.

Environmental profile

VCIPL assumes environmental responsibility towards its employees and the community through a structured EMS (environment management system) and OHSMS (occupational health and safety management system) that is designed to meet the requirements of ISO 14001: 2004 and OHSAS 18001: 1999.

VCIPL has also taken initiatives by recycling solvents, reusing boxes, reducing the use of packaging material, and conserving energy and water. It also provides facilities for employees to undergo career-counselling sessions, provides money for scholarships for the children of employees, and organizes a summer camp for the children. The company participates in blood donation camps, and provides earthquake and drought relief funds.

The main air pollutants emitted by VCIPL are SPM (suspended particulate matter), RSPM (respirable suspended particulate matter), SO_2 (sulphur dioxide), NO_x (oxides of nitrogen), and noise. The SPM, SO_2, NO_x, and CO_2 (carbon dioxide) from the diesel generator stack are dispersed at an appropriate height, while RSPM and SPM from the powder coating in the film cap and laser room are not dispersed into the atmosphere. In order to control the noise pollution, the company uses a noise level controlling system.

The main water pollutants discharged by VCIPL are BOD (biochemical oxygen demand), COD (chemical oxygen demand), TDS (total dissolved solids), TSS (total suspended solids), pH, and oil and grease. In order to control all forms of water pollution and effluent, the company has developed, established, and implemented a waste disposal system and an industrial ETP (effluent treatment plant). Further, as part of its water conservation programme, the company uses treated water for gardening purposes.

The hazardous waste generated by VCIPL is collected in different containers and segregated into various categories. The entire quantity of waste is sent to Mumbai Waste Management Ltd, a common hazardous waste disposal site that is approved by the MPCB (Maharashtra Pollution Control Board).

Case Study

Electronic components have diverse uses in the consumer electronics and general industrial sectors. However, they also have various adverse impacts on the environment. VCIPL produces a vast range of resistors and capacitors for its customers, and is sensitive to the changing needs of the environment in which it is applied.

Virtually, the entire range of these products use solderable coatings. The key ingredients used to make these coatings effective are Pb (lead) and Sn (tin), in various proportions. However, studies have shown that Pb has toxic properties and it is classified by the EPA (Environmental Protection Act) as a 'probable human carcinogen of class B2'. The harmful effects that result from it are water pollution, impact on the central nervous system poisoning, etc. Pb is known to cause irreversible brain damage, haematological anemia, gastrointestinal colic, and kidney and thyroid malfunction.

VCIPL has taken proactive measures to tackle this issue, and has launched a factory-wide initiative to eliminate the use of Pb from all its manufactured products.

The measures implemented include finding suitable suppliers of Pb-free termination wires. The manufacturing supplier sites were visited and assessed, and sample wires were also tested. Test and verification measures followed the criteria used by Sony-Japan and Vishay Germany, in the absence of IEC (International Electrotechnical Commission) or comparable standards. Facilities necessary for testing were also set up. Multilevel results were obtained from widespread test laboratories; these were consolidated using an array of test facilities, and were used to evolve training programmes and standard operating procedures, and to identify test material suppliers who could meet the requirements.

VCIPL also made several modifications to the equipment and processing conditions so that they were capable of handling Pb-free termination wires. The various processes included the following.

- Specifying packaging and transportation methods
- Storage and inspection of incoming goods
- Handling, use, and application on the production shop floors
- Fulfillment of acceptance criteria for finished goods
- Zero hour and reliability test standards
- Review of shelf life control measures

Further, VCIPL opened its manufacturing units for audits by customers and external agencies. Access process for customers and certification bodies was simplified. Other measures include standardizing work methods, training and educating the workforce, and responding to customer requests for technical assistance.

The case study led to a reduction in the annual consumption of 284 tonnes of Pb-contaminated termination wires—the entire quantity was replaced by non-Pb-contaminated termination wires. As a result, there has been a reduction in pollution and in-process rejections, and improvement in processing efficiency. This is also beneficial to employees as they now work in a Pb-free environment and do not suffer from the ill effects that arise out of handling Pb-contaminated parts.

Sony-Japan, having pioneered the green manufacturing idea, was invited by VCIPL to audit the site for verification after completion of the Green Electronic Components Project, and has been certified as a 'green-partner' of Sony-Japan.

CASE STUDY

at a glance

- VCIPL (Vishay Components India Pvt. Ltd) is one of the world's largest manufacturers of discrete semiconductors and passive components.
- VCIPL produces a vast range of resistors and capacitors for its customers. Virtually the entire range of these products have coatings on them, which are needed when they are soldered in electronic circuits.
- The important ingredients in making these coatings effective are Pb (lead) and Sn (tin).
- In order to stop using Pb, VCIPL found suitable suppliers of Pb-free termination wires and set up facilities that were necessary for testing.
- As a result of using non-Pb-contaminated wires, pollution and in-process rejections have reduced, and processing efficiency has improved.

Environment Management System

CASE STUDY

22

Infosys Technologies Ltd

Ozone initiative: the Infosys health, safety, and environmental system

Background

Infosys Technologies Ltd is a global provider of consulting and IT (information technology) services. The company has over 72 000 employees worldwide and had an annual turnover of over Rs 91.7 billion as on 31 March 2006. The company's Infosys campus is located in Bangalore.

Environmental profile

Infosys has a formalized EMS (environmental management system) and is certified to ISO-14001 at eight development centres in India.

The company takes various initiatives towards achieving resource conservation, pollution prevention, effective waste management, adherence to all applicable legislation, and awareness creation among employees. In terms of social responsibility, Infosys strives to create awareness amongst employees; donates ambulances, traffic control vehicles, traffic lights, road signs, timers, etc. for traffic management; works in the areas of health care, education, rural upliftment, and arts to support the underprivileged; extends support in time of natural disaster or calamities; and sponsors community improvement initiatives.

Infosys has a number of measures in place to mitigate air pollution. Flue gas, which consists of pollutants like SPM (suspended particulate matter), SO_2 (sulphur dioxide), and NO_x (oxides of nitrogen), as a result of use of DGs (diesel generators), is mitigated by keeping the chimney height as specified by KSPCB (Karnataka State Pollution Control Board) norms, thus ensuring proper dispersion of DG exhaust. Noise pollution caused by the use of the DG sets is mitigated by acoustically housing the generators, and the personnel working in the vicinity are provided with protective equipment like ear-muffs.

Water pollutants discharged include BOD (biochemical oxygen demand), COD (chemical oxygen demand), pH, oil and grease, TDS (total dissolved solids), and TSS (total suspended solids). Abatement of water pollutants is achieved at Infosys through its three-stage sewage treatment methodology, after which it is used for irrigation purposes.

Hazardous waste generation at Infosys is at a minimal level. Used oil, cotton waste with oil/grease, filters, etc. are hazardous wastes generated by the DG sets and are stored and disposed off to a vendor approved by the state pollution control board and the CPCB (Central Pollution Control Board). E-waste is stored as per the norms and will also be disposed off to external vendors.

The facility has an environmental risk management and disaster management plan in place.

Case study

The Ozone Initiative

The 'Ozone Initiative' is the EMS at Infosys, which helps the company comply fully with all the legal requirements or exceed these expectations. It drives Infosys to be an environment-friendly company with elective systems of waste management, recycling, etc., which helps in the abatement of pollution and conservation of natural resources. It also help provide a secure working environment to protect its employees.

A number of initiatives have been undertaken at Infosys as a part of the Ozone Initiative, including the following.

- Installation of VFDs (variable frequency drives) for AHUs (air handling units) to reduce energy consumption, and installation of timers in AHU motors for controlling operational hours
- Impeller trimming of chilled water and condenser water pumps to reduce energy consumption
- Introduction of sensor taps in food courts and urinal sensors to avoid wastage of water
- Installation of dishwashers in food courts for water conservation
- Recycling of waste paper
- Introduction of vermiculture on the campus

This case study focuses on two such projects

Project 1: Energy conservation at the multi-level car parking lot

The Infosys campus, catering to a large number of visitors and employees in the campus on a daily basis, established an MLPL (multi-level parking lot). The MLPL was constructed to reduce congestion on the internal roads where cars were parked earlier.

In order to conserve energy at the MLPL and, in turn, reduce costs to the company, timers were introduced to control the switching on/off of lights. All timers were installed and set to operate at two different timings with contractor logic. All lights were programmed to come on at 6:45 p.m. and switch off at 11:30 p.m., and very few lights were kept on till the morning. The delayed setting of timing by floor ensured that all floor lights were not switched on at the same time. Thus, energy consumption was reduced from 25 000 units per month to 15 000 units per month. This also led to a total savings of about Rs 840 000 per annum. Figure 1 shows the energy consumption data for the Infosys MLPL, from October 2005 to October 2006.

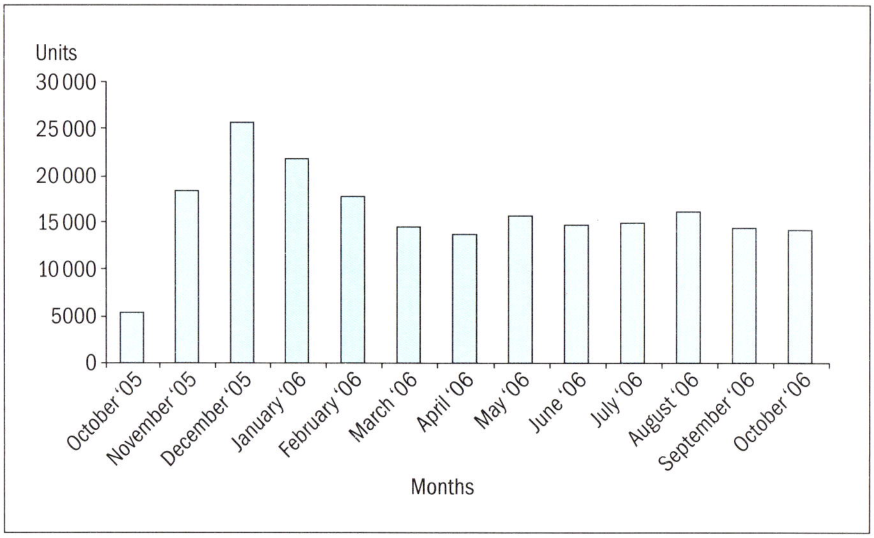

Figure 1
Energy consumption at the multi-level parking lot

Project 2: Installation of solar water heating system at the Employee Care Centre

In order to accommodate employees in transit, Infosys houses an ECC (Employee Care Centre) on its campus. The ECC resembles a guest house in terms of facilities available. The need for the ECC to be operational throughout the day, everyday of the year, implied room to reap benefits from harnessing solar energy.

In order to enable water heating for the increased number of personnel staying at the ECC, Infosys set up a solar water heating system as shown in Figure 2. Conventional heating required the use of 168 kW (kilowatt) heaters to heat 500 LPD (litres per day) of water in Bangalore's environment. Thus, it was estimated that through the conventional system, the ECC would require about 1.47 million units of energy per annum. The 168-kW capacity conventional heaters remain installed at the ECC only as a back-up, in case of absence of solar radiation.

The installation of the solar water heating system at the ECC resulted in a saving in energy consumption of about 1.47 units per annum and in cost of Rs 7.20 million. This system has further been enabled at all the company's development centres across India to a capacity of 15 000 LPD, resulting in savings of 2.31 units in energy consumption and Rs 11.56 in cost, for a period of 300 days.

Figure 2
Solar water heating system at Employee Care Centre

CASE STUDY

at a glance

- Infosys Technologies Ltd, Bangalore, a global provider of consulting and IT (information technology) services, had a turnover of over Rs 91.7 billion, as on 31 March 2006.
- The 'Ozone Initiative' is the EMS (environmental management system) at Infosys, which drives the organization to be environment friendly by managing waste, recycling, abating pollution, conserving natural resources, etc.
- Two projects under the Ozone Initiative entail the conservation of energy at the campus MLPL (multi-level parking lot) and the installation of a solar water heating system at the ECC (Employee Care Centre).
- Under the first project, timers were installed at the MLPL to control the switching on/off of lights.
- This project resulted in a reduction of energy consumption from 25 000 units per month to 15 000 units per month, and an annual cost saving of about Rs 840 000.
- The second project entailed the setting up of a solar heating system at the ECC, thus saving 1.47 units of energy per annum, which was initially consumed by the use of 165-kW capacity conventional heaters and a cost saving of Rs 7.20 million.

CASE STUDY

23

JK Paper Mills Unit, JK Paper Ltd

Measures taken for enhancing environmental performance

Background

The JKPM (JK Paper Mills) unit of JK Paper Ltd is the flagship company of the JK organization, and is an integrated pulp and paper mill. The mill was set up in 1962 at Jaykaypur, in the district of Rayagada, Orissa, and has an installed capacity of 18 000 TPA (tonnes per annum). An aerial view of the mill is given in Figure 1. Its turnover for 2005/06 was Rs 5950 million. JKPM manufactures paper and paper board, pulp-drying sheets, and coated paper and board. Annual production for the same period was 126 262 MT (million tonnes).

Environmental profile

JKPM has a strong commitment to continuously improving its performance in the area of SHE (safety, health, and environment). The efforts and practices adopted by the mill have made it the first paper mill in India to be awarded with the ISO-9001: 2000 QMS (quality management system) certification and ISO-14001: 1996 EMS (environmental management system). It has also won a host of other accolades including Greenest Paper Mill Award of CSE (Centre for Science and Environment) in 1999, Sword of Honour from the British Safety Council, and second best Pollution Control Excellence Award for implementing effective pollution control measures in 2004.

JKPM takes measures to abate or mitigate air pollutants. ESPs (electrostatic precipitators) are installed to arrest SPM (suspended particulate matter), and there is a coal, lime dust extraction system for fugitive emissions. Furthermore, conversion of pulping from an open blowing to a closed system through technology change and an investment of Rs 3 billion at the pulp mill have led to the successful arrest of odour. Better combustion burners in boilers reduce SO_2 (sulphur dioxide) and NO_x (oxides of nitrogen).

Modern activated sludge processes at the ETP (effluent treatment plant) reduce BOD (biochemical oxygen demand) and COD (chemical

Figure 1
Aerial view of JK Paper Mills

oxygen demand). There is a neutralization system for pH, cooling tower for cooling the effluent temperature to 34–37 °C, skimmer arrangement at the ETP for oil and grease, and primary and secondary clarifiers in the ETP to contain TDS (total dissolved solids) and TSS (total suspended solids).

There are effective disposal measures for the reduction or recycling of hazardous waste. Used oil due to maintenance is sold to outside authorized parties for reprocessing, spent resin from the DM (demineralized) plant is burned in the boilers (JKPM is in the process of developing a scientific landfill system), and oil consumption is decreased by better maintenance through TPM (total productive maintenance).

JKPM is in the process of implementing a CDM (clean development mechanism) project in consultation with the World Bank for three districts in both Orissa and Andhra Pradesh. The project is titled 'Improving rural livelihoods through carbon sequesteration by adopting environment-friendly technology-based agro forestry practices'.

Case study

Measures to improve environmental performance include afforestation through social and farm forestry supported by clonal technology, cleaner technologies and processes, reducing pollutants in discharged water, reducing particulate emissions, solid waste management, and conservation of resources.

Greening and development of wastelands and high yield pulpable species

In 1990, JKPM established its plantation wing for the generation of raw material in participation with farmers with the following broad objectives.

- Creating plantation awareness and establishing of plantations with peoples' participation by utilizing their dry/wastelands.
- Augmenting raw material availability with a view to self-sustainability.
- Improving socio-economic conditions through the generation of rural employment.
- Environment rehabilitation.

JKPM provides high-quality saplings through a network of decentralized nurseries, technical know-how, and buy-back arrangements for the harvested pulpwood through social and farm forestry. Table 1 shows area covered (by state) and number of saplings and clones distributed from 1990 to 2006.

Measures in the process

Chipping, pulping, and bleaching process

Air cyclones to separate dust, and belt conveyors for reducing dust have been provided at the chipper. Digester relief, and acidic and alkali vent gases emitted during cooking and bleaching are arrested through a scrubber. An auto chemical spill collection system was also installed. As a result, JKPM achieved an AOX (absorable organic halides) level of 0.4–0.8 kg per tonne of paper, as shown in Figure 2, which is far lower than any other pulp and paper industry in India, and within the regulatory limits.

Chemical recovery

Condensers have been provided to condense vapour generated in the plant and prevent heat from going into the atmosphere. Cascade evaporators have been eliminated by installing high-surface-area economizer in the LFB (liquor-fired boiler) to reduce emission and achieve higher steam

TABLE 1
ACHIEVEMENTS IN PLANTATION (1990–2006)

Plantation year	*Area covered (by state)* Orissa	*Andhra Pradesh*	*Chhattisgarh*	*West Bengal*	*Total area covered (Rs)*	*Number of saplings and clones distributed (million)*
1990	55	–	–	–	55	0.1
1991	233	–	–	–	233	0.6
1992	187	–	–	–	187	0.5
1993	254	–	–	–	254	1.1
1994	733	–	–	–	733	2.0
1995	1267				1267	3.4
1996	2561	199	–	–	2759	8.8
1997	3140	713	–	–	3853	10.3
1998	3508	726	–	–	4234	10.6
1999	1760	928	–	–	2688	8.4
2000	1896	863	–	–	2759	12.6
2001	2426	1899	–	–	4325	11.6
2002	1676	2329			4005	25.1
2003	1238	2297	64	–	3599	22.1
2004	1525	2590	52	64	4231	26.3
2005	1422	3034	78	111	4645	29.0
2006	1483	2851	50	106	4490	28.5
Total	25 364	18 429	244	281	44 317	201.0

generation. Dust extraction systems for lime handling, efficient ESPs to arrest solid particles from going into the atmosphere, tall stacks and higher exit velocities, scrubbers for absorbing vent gasses of the smelt-dissolving tank, and an auto-chemical spill-collection system have also been installed. As a result, JKPM has achieved the highest chemical recovery percentage (>95%) in India, resulting in significant improvements in discharged effluent and air quality (Figure 3).

Stock preparation and paper machines

Here, paper machine backwater is reused after clarification in pulp mill and other areas. Effective centricleaner system to reduce fibre loss and

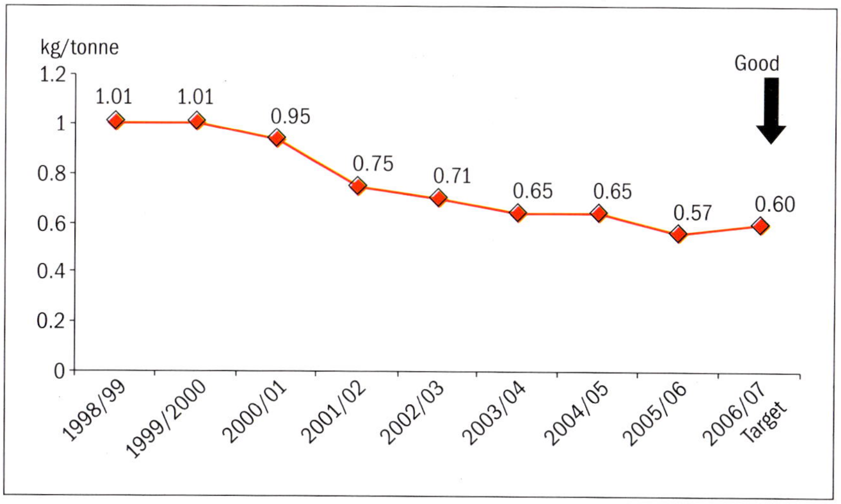

Figure 2
AOX level at JKPM (1998–2006)

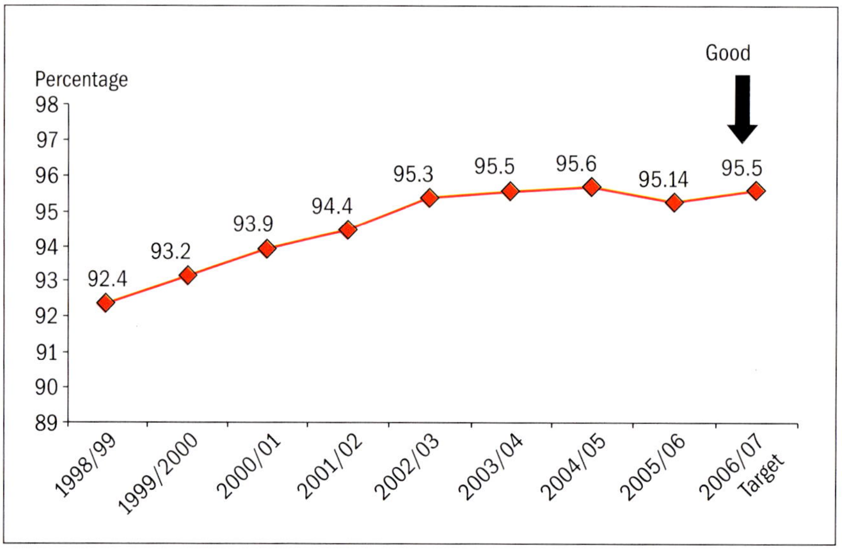

Figure 3
Chemical recovery (%) (1998-2006)

'save alls' in paper machines have also been installed. Clarified water is completely recycled in wire cleaning, centricleaner dilution.

Utility

All CFBs (coal-fired boilers) have been provided with efficient ESPs with tall stacks and higher exit velocities. A dust extraction system (dry fog system) has been installed for the coal-handling system.

Effluent treatment plant

JKPM upgraded its existing ETP (effluent treatment plant) by incorporating an advanced activated sludge system with a cooling tower for further improvement in the quality of effluent (Figure 4). The upgraded cooling tower in the ETP maintains the temperature profile of the effluent before it goes to the aeration basin. The optimum temperature would enhance the rate of growth of microorganisms, which eat up most of the pollutants. Such an aeration basin with a cooling tower in the ETP is the first of its kind in the Indian pulp and paper industries.

Figures 5a and 5b show the resulting fall in BOD and SS (suspended solids) from 1998 to 2006.

Figure 4
View of the state-of the-art effluent treatment plant at JKPM

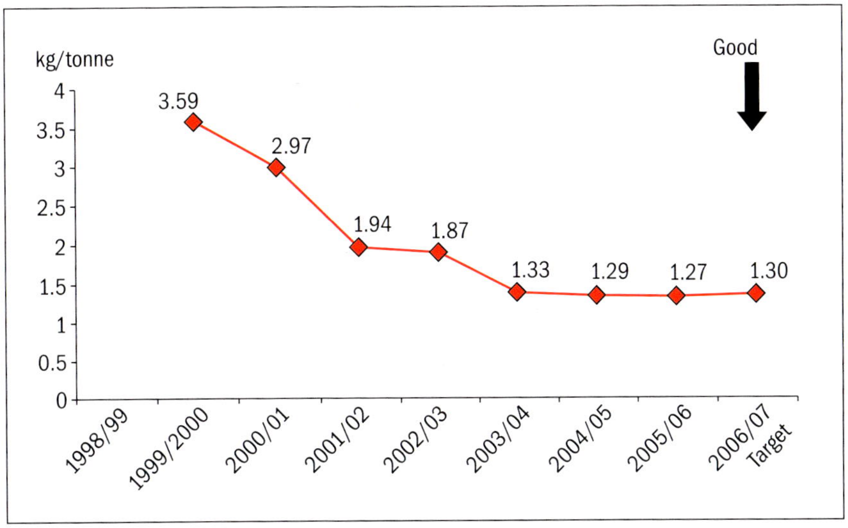

Figure 5a
Biochemical oxygen demand (kg per tonne of paper) (1998–2006)

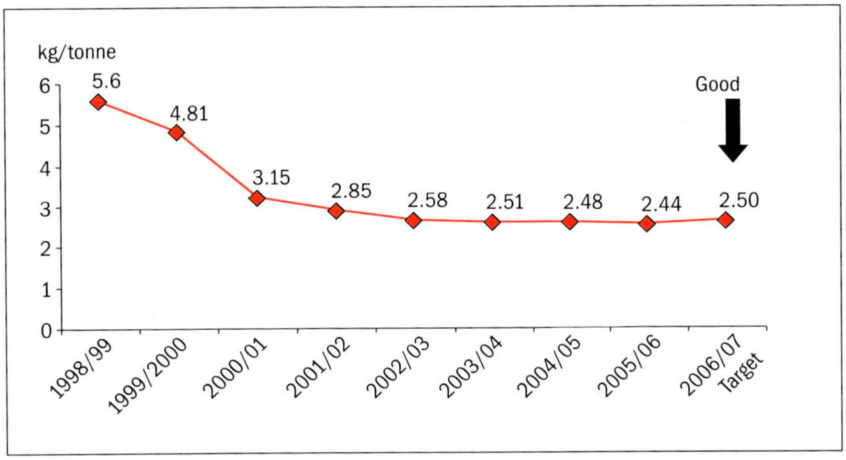

Figure 5b
Suspended soild (kg per tonne of paper (1998–2006)

Solid waste management

JKPM successfully utilizes different types of solid wastes generated by it. Fly ash, for instance, has good pozzolanic properties and is a mineral admixture inert material, thus used for filling of low-lying areas, brick manufacturing, and road sub-base making. Figure 6 shows the percentage of fly ash utilization by JKPM from 1998 to 2006.

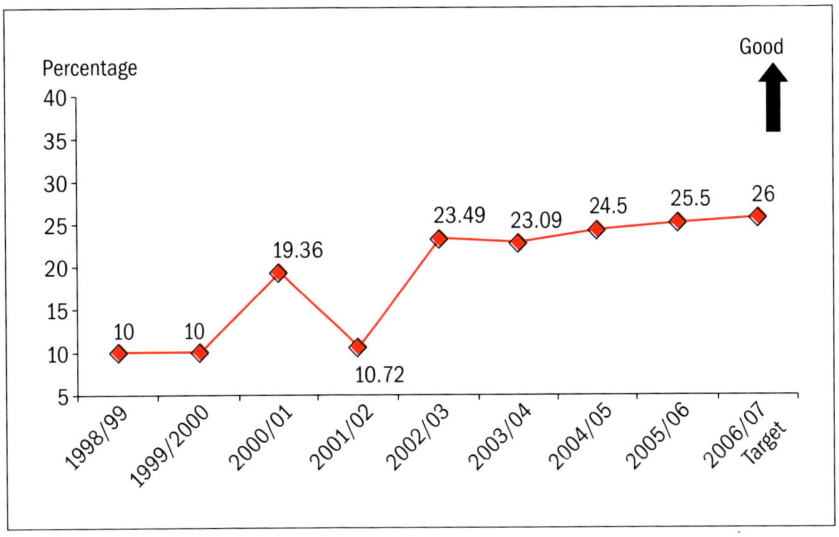

Figure 6
Fly ash utilization (1998–2006)

Lime sludge is also utilized for the filling of low-lying areas, land-filling, and planning for reuse in the lime kiln. ETP primary sludge is used for the manufacture of low-grade paper products, while ETP bio-sludge which is rich in NPK (nitrogen, phosphorus, and potassium) is used as manure by local farmers. Bamboo and hard wood dust are used as fuel for CFBs.

Energy conservation

JKPM has implemented a number of projects for resource conservation. Energy conservation is an area of importance at the mill. While expansion at the mill could overburden the grid demand, JKPM has set up its own additional power generation plants and modernized/discarded its old boilers and pulp mills. JKPM has an energy conservation committee and policy, which aim to minimize losses and wastage, use techniques to substitute high-cost energy by low-cost energy, optimize consumption, and maximize utilization of plan and machinery capacity. These goals are achieved by undertaking studies on energy consumption trends in terms of power, steam, and water in various sections at the mill; following up with various departments to arrest steam leakages, condensate leakages, water leakages; providing adequate insulation required to minimize heat losses; engaging external agencies to assess the health of the organization with regard to energy consumption; optimizing investment for additional power generation station; encouraging consciousness among employees about energy scarcity and inculcating the habit of conserving and effectively using energy; imparting training to improve the skills of personnel to evaluate the opportunity by material balance, heat balance, energy balance, etc.

CASE STUDY at a glance

- JKPM (JK Paper Mills) is a unit of JK Paper Ltd that has an installed capacity of 18 000 TPA (tonnes per annum) of pulp and paper board, pulp-drying sheets, and coated paper and board.
- The environment policy at JKPM aims to comply with applicable environmental legislations, prevent pollution, and continually improve environmental performance.
- The mill has taken a number of steps to enhance environmental performance.
- JKPM is involved in afforestation through social and farm forestry supported by clonal technology.
- Cleaner technologies and processes are developed at the mill in, for instance, chipping, pulping, and bleaching process; chemical recovery; and stock preparation and paper machines.
- Pollutants are reduced in discharged water by ETPs (effluent treatment plants).
- Particulate emissions are reduced by various means.
- Solid waste management is undertaken at the mill whereby fly ash, lime sludge, ETP primary sludge, ETP bio-sludge, and bamboo and hard wood dust are effectively utilized.
- Resource conservation is another area of importance, wherein energy is conserved .

CASE STUDY

24

National Hydroelectric Power Corporation Ltd

Implementation of Environmental Management System at Uri Power Station: A Stride Towards Sustainable Development

Background

NHPC (National Hydroelectric Power Corporation Ltd), a Government of India enterprise, was incorporated in 1975. NHPC's Uri Power Station at District Baramulla, in the state of Jammu and Kashmir, produces clean and environment-friendly hydro power, and has an installed capacity of 480 MW (megawatts). It is situated on the Jhelum river, shown in Figure 1.

Environmental profile

As hydro power projects use renewable and non-polluting resources for power generation, no polluting gases are emitted, unlike power generation with fossil fuels. At the Uri Power Station, no water pollutants or hazardous wastes are generated.

NHPC is an ISO 14001:2004 certified company with an environmental policy and a department at both the corporate and facility levels. The environment policy adopted by NHPC aims at addressing the environmental concerns in the sector of hydro power development, as well as maintaining and continually improving the EMS (environmental management system) to conform to international standards. NHPC has always conformed to all the legal requirements specified under the Environment (Protection) Act, 1986; Forest (Conservation) Act, 1980; and Wildlife (Protection) Act, 1972; and has shown continued commitment towards CSR (corporate social responsibility).

NHPC's CSR initiatives include providing educational facilities in the area in and around project sites and promoting arts, crafts, sports, and other cultural activities to deserving persons. Apart from the NHPC project schools, the opening of Kendriya Vidyalaya schools is also encouraged by provision of basic infrastructural facilities like buildings,

Figure 1
NHPC Barrage Complex at Uri Power Station at Baramulla, Jammu and Kashmir

electricity, water supply, furniture, etc. Programmes are in place such as the Adult Education Programme, and vocational training courses for men, women, and people with disabilities for their economic upliftment. Training programmes for improving the condition of women at various project sites include teaching them stitching; crafting; processing of pickles, jams, jelly, etc. Aid has been given to an orphanage in the vicinity to purchase bunk beds, stationary, shoes, and other basic amenities.

Further, eye camps are organized regularly in which operations are carried out and spectacles are distributed. Financial assistance is provided to the old and needy to undergo urgent operations or pay for treatment. At all its project locations, NHPC has set up well-equipped project hospitals (complete with biochemical analysers, x-ray machines, ECG machines, labour room, and operation theatres for minor surgeries). These hospitals provide free consultations for local residents. Comprehensive medical check-up camps, vaccination programmes, and awareness programmes on various communicable and vector-borne diseases are organized. Medication is provided and mother–child kits consisting of clothes, oil, soaps, biscuits, medication, etc. are distributed.

A number of rain shelters have been constructed for locals in hilly regions that receive substantial rainfall. Hand-pumps for water have also been installed. Dish antennas have been installed for direct reception of television programmes, and colour television sets have been distributed free of cost to some villagers.

For employees at NHPC, safety norms have been clearly laid out and all possible measures are taken to keep the workforce safe from fire, natural disaster, emergency situations, etc. There is a disaster management team that responds to calamities, and organizes relief and rehabilitation. For example, after the earthquake of October 2005 at Jammu and Kashmir, NPHC adopted six villages in the state to carry out relief and rehabilitation work.

Case study

NHPC has identified environmental conservation as a cross-cutting issue in the overarching framework of sustainable development. The Uri Power Station is a project of considerable magnitude, and huge civil and electrical structures have been constructed for the generation of power. Thus, action is taken to ensure that pollution is mitigated or minimized from all project activities.

ISO 14001: 2004 certification further pushed the Uri Power Station to streamline EMS (environmental management system) at the project, which could, in turn, help in generating power in an environmentally sustainable manner. The power station thus can project itself before the national and international community as an environmentally conscious power station.

The Uri Power Station has various EMPs (environmental management plans) in place to address energy conservation, oil conservation, disposal of garbage, disposal of biomedical waste, construction of oil catcher, phasing out of CFC (chloroflouro carbon) gas, afforestation, etc. to achieve the above-mentioned goals.

Construction and operation of the power station

In the execution of the Uri Hydroelectric Project, the most modern construction techniques were used such as the latest type of scaffolding, cable anchors for the power house walls, and raise borer/raise climbers for shaft excavation.

In order to achieve the objective of displacing less efficient and polluting forms of power generation, the most sophisticated and modern equipment is utilized in the Uri Power Station and their operation is fully computerized. The power plant is operated by highly trained and skilled engineers, and there is a well-qualified environmental monitoring team to monitor all aspects of running the project.

Resettlement and rehabilitation of affected population

The Uri Power Station, in association with the State Revenue Department and representatives of the local population, conducted a detailed socio-economic survey and formulated an R&R (resettlement and rehabilitation) plan for project-affected persons. A broad R&R package was developed, consisting of compensation for land, houses, shops, and other properties; transportation charges for household items, cattle, etc.; construction of houses; solatium charges; financial assistance for construction of cattle sheds or poultry farms; agriculture land, depending on availability; subsidies for seeds, fertilizers, and land management; development of public health centres, schools, community centres, etc; basic amenities like roads, drinking water, electricity, medical facilities, etc.; renovation or relocation of religious structures; and so on. Figure 2 shows reconstructed houses at Uri.

Waste disposal

Although hydro-power projects are renewable and non-polluting sources of power generation, releasing no air or water pollutants, some pollution from day-to-day activities of the population residing in the project area does take place. The waste generated from the offices, colonies, canteens, workshops, etc. has the potential of polluting land and water resources. EMPs are in place for arresting these pollutants. Garbage is segregated into biodegradable and non-biodegradable forms and stored in designated bins. The biodegradable waste stored in the bins is fully disposed off at a designated garbage-dumping site.

An oil catcher is constructed to reduce the contents of oil and grease from waste water generated from the workshop. Similarly, proper EMPs for the disposal of biomedical and hospital waste, waste oil, and batteries, and prevention of oil leakage are also being implemented at the project.

Figure 2
Reconstructed houses at Uri

Catchment area treatment plan

Initially, the misuse of land and its resources had caused considerable damage to the catchment of the project. The area was also highly vulnerable to erosion; biotic interference in the forest lands, encroachment, and heavy grazing, coupled with weak geological formations and rugged topography, had aggravated soil erosion problems in the area. This resulted in deep gully formation, landslides, loss of topsoil cover, with consequent high yield of sedimentation. Therefore, intensive soil conservation measures were taken to provide maximum possible cover to the land surface, and intensive land development and torrent control measures were taken in the catchment.

To arrest soil erosion from the catchment, which could encroach upon the efficiency of the project, a CAT (catchment area treatment) plan was prepared for 13 180 hectares of the catchment area, covering nine micro-watersheds. Under this plan, various measures were implemented including bench terracing and contour bunding; gully control measures like check dams and crate works and landslide control measures like retaining walls, fascine structures, and gabions. In addition, large-scale plantation works (Figure 3) were also undertaken.

Restoration of muck dumping sites

The construction of various open and underground structures required excavation of the landscape of the project area, which, in turn, generated approximately 5 500 000 m^3 (cubic metres) of muck that was dumped at 10 designated locations. In association with NEERI (National Environment Engineering and Research Institute) of Nagpur, an integrated biotechnological approach to rectify 95 hectares of muck-dumping sites was adopted for the first time in the state of Jammu and Kashmir. Luxuriant tree cover was provided to strengthen and stabilize the dumping surface, and has, thereby, prevented spilling over of muck either on land or in the Jhelum river as shown in Figure 4. Suitable plant species of high ecological and economic value adaptable to local conditions were planted at the rate of 1800–2200 saplings per hectare at spoil tips—namely, *Populus* sp., *Salix alba*, *Robinia pseudoacacia*, *Ailanthus excelsa*, and *Cupressus* sp. Waste material dumped at spoil tips, comprising mainly loose rock fragments, had been mechanically compacted and properly levelled with suitable slopes.

Figure 3
Population in catchment area treatment initiatives

Figure 4
Muck dumping sites after restoration

Afforestation

The project entailed diversion of 54.6 hectares of forest land for its construction activities and the felling of about 4000 trees. Thus, compensatory afforestation of 156 000 trees over 62.7 hectares of degraded forest land was carried out. An additional 320 000 trees were also planted. The afforested area was fenced and fast-growing trees useful for fodder and fuelwood were planted (Figure 5). About 35 000 shade-providing and fruit-bearing species were also planted.

Construction of fishway

Under the aquatic environment assessment programme, the Uri Station provided a fishway across the Uri barrage for the movement of fish, particularly snow trout, most species of which are migratory in nature and migrate upstream for spawning. A fish ladder was also constructed over Buniyar *nullah*, a tributary of the Jhelum river, for facilitating the movement of migratory fish species, shown in Figure 6.

Figure 5
Afforestation site

Figure 6
Fish ladder constructed over the Buniyar *nullah*

Energy conservation

Several measures are rigorously followed at the Uri Power Station to ensure energy conservation. All employees are instructed to put off lighting, heaters, and coolers during the daytime and use air conditioners only when required. Only 28- or 36-watt tube-lights or CFLs (compact fluorescent lamps) are used, and reflectors in the light fittings are regularly cleaned. Periodic maintenance to ensure all power factor correcting panels and temperature controllers are functioning satisfactorily. Water pumps are also used judiciously. Compressor operation is controlled using pressure switches, and temperature controllers are used in chillers.

CASE STUDY

at a glance

- NHPC (National Hydroelectric Power Corporation Ltd) has a hydro power project, the Uri Power Station, with an installed capacity of 480 MW (megawatts) on the Jhelum river in Baramulla, Jammu and Kashmir.
- The Uri Power Station is a renewable and non-polluting source of power generation, releasing no air or water pollutants, and the facility is committed to CSR (corporate social responsibility) activities and environmental conservation, generating power in an environmentally and socially sustainable manner.
- The goal of the environmental policy of the Uri Power Station is to continually improve its EMS (environmental management system) to conform to international standards, in line with its ISO 14001: 2004 certification.
- In order to maintain and continually improve its EMS, the Uri Power Station has put various EMPs (environmental management plans) in place that are rigorously implemented.
- EMPs with respect to catchment area treatment, compensatory afforestation, restoration of muck disposal sites and construction areas, and resettlement and rehabilitation were implemented to address the anticipated environmental and social risks ensuing from the construction activities.
- A significant initiative under an EMP, in association with NEERI (National Environment Engineering and Research Institute), Nagpur, was the use of an innovative biotechnological approach for the restoration of 95 hectares of muck dumping sites where approximately 5.5 million metres of excavated material was dumped.

CASE STUDY

25

Wipro Technologies Ltd

Wipro Gurgaon Software Development Centre: A Green Building

Background

Wipro Technologies Ltd began its operations about 60 years ago as a company manufacturing vegetable oil and is, today, one of the world's leading technology service providers. The year 2005/06 has witnessed revenues in excess of Rs 100 billion, profit after tax of over Rs 20 billion, and an increase in the company's team size to over 50 000. The presented case study is based on the company's software development facility in Gurgaon, Haryana (Figure 1).

Environmental profile

While Wipro does not have an environment department, it has an environment policy at the facility level and ISO 14001 certification through its initiatives towards energy and water conservation, waste management, legal compliance, and ensuring safety at the work place. Noise and emissions are regularly monitored on a monthly basis, and it is ensured that they are kept within the PCB (pollution control board) norms. There is an STP (sewage treatment plant), and treated water quality is regularly monitored to ensure that it falls within the PCB's prescribed norms. Waste oil and oil-soaked cotton generated from the DG (diesel generator) are sent to the CPCB (Central Pollution Control Board)–authorized vendors for recycling and incineration. E-waste is disposed off to E-Parisara, approved by the CPCB.

A number of other initiatives taken by Wipro to set an example in environmental leadership can be seen in the design and construction of the building itself. Energy conservation, water conservation, and waste management through its 'green building' are further discussed in the case study.

Figure 1
Wipro Gurgaon Software Development Centre

Case study

Wipro, Gurgaon, is located in an area dedicated for industrial and community buildings, and hence, the site is subjected to high levels of pollution. Several measures have been taken in the construction of the centre, as well as in its daily running, that have led Wipro's 'green building', its Gurgaon Software Development Centre, to join the elite club of LEED (Leadership in Energy and Environmental Design)—certified Platinum-rated buildings worldwide, awarded by the US Green Building Council. Moreover, value-for-money solutions have been implemented to reduce its environmental footprint—Wipro has achieved significant savings on recurring costs (of about 30%) with only a nominal increase in capital costs (10%).

The building

The Gurgaon Software Development Centre is built on a plot of 1.12 acres, with a built-up area of 181 000 ft^2 (square feet). The building comprises seven floors, two basements, and two terraces. Total air-conditioned area is 90 000 ft^2, and the total connected load is 1300 kVA (kilovolt amperes).

In constructing the building, Wipro was careful to preserve and maintain the existing vegetation and ground covers. All the existing trees along the periphery were retained to stabilize the soil and minimize erosion potential. This has also helped in rainwater infiltration, thereby reducing the quantity and velocity of storm water run-off.

Further, the excavation area for the double basement was minimized to avoid unnecessary site disturbance and erosion. To prevent soil erosion from the vertical faces of the excavated area, five metre long MS (mild steel) rods were driven into the soil after every one metre of distance. The topsoil, rich in organic content, was stripped and stacked for further use, and the turf from the excavated area was transplanted at the nursery for future use. Geotextile-like non-woven geofabric was used over the plants on the basement slab, over the filter layer, and for the protection of perforated pipes. This prevents fine soil particles from entering into the pipes and clogging them.

The building has made maximum use of environment-friendly and recyclable products such as fly-ash-based ACC (aerated concrete) blocks, recycled wood certified by the Forest Steward Council, and ceramic tiles. It sources about 40% of building material from within 500 miles of the project site, and uses modular furniture made from steel-powder-coated tiles and steel fabric tiles, which are emission-free and long lasting.

Water efficiency

Wipro conserves water through a number of effective steps. One such step is the implementation of water-efficient landscaping. Pop-up sprinklers are used for roof turfs and turf surfaces while the drip system is used for all shrubs, ground covers, and trees. The entire water requirement for irrigation is met through treated water. Soak pits and wells have been constructed along the perimeter of the site to harvest rainwater run-off from roof tops and hard areas.

An STP with a capacity of 70 000 litres per day has been set up to recover the effluent of quality, suitable for use as make-up water for cooling towers of the air-conditioning system, landscaping, and flushing. The process adopted is a high-efficiency aerobic-treatment scheme, based on the extended aeration system, and is capable of withstanding shock loads. The water for the cooling towers is further passed through a softener.

Water meters have been installed at different usage points to monitor consumption.

Energy conservation

Considering Gurgaon's hot and dry climate, special emphasis was laid on designing an efficient air-conditioning system, with a focus on energy conservation.

Wipro divided its office spaces into different zones, based on occupancy pattern and ease of air distribution. Each zone was provided with an individual AHU (air-handling unit). The cooling coils of all AHUs are provided with motorized automatic controls and balancing valves. Space temperature and relative humidity monitoring for reach zone is carried out through sensors located in the return air stream of each AHU. These sensors report to an advanced building management system, which, in turn, controls the motorized control valves, electric tubular heaters, and humidifier, thereby achieving precise control over temperature and humidity. This building automation system not only helps in conserving energy by making it possible to plan and execute various conservation and control schemes, but also helps in reducing trained manpower requirement for operating and maintaining the building services. The air quality in each space is closely monitored through CO_2 (carbon dioxide) sensors installed at each AHU.

The walls of the building were constructed using ACC blocks that have a definite shape, and are light in weight, environment-friendly, and manufactured using recycled material. Double glazed windows with energy-efficient glass ensure not only low noise levels, but also low penetration of heat into the building, thereby reducing the load on the air-conditioning system and resulting in lower energy consumption.

The glass specification for windows enhances the availability of daylight in regularly occupied spaces, and maintains visual comfort for the occupants while proving to be energy efficient. The 'light well' at the centre of the building, shown in Figure 2, is an architectural feature that

Figure 2
Light well at Gurgaon Software Development Centre

helps light percolate into the offices spaces. Daylight glazing areas reduce with the increase in the height of the building. The building thus saves nearly 49% of lighting energy over a budget building—the average LPD (light power density) for a budget building is 1.28 W (watts)/ft^2, while the average LPD achieved at the centre is 0.8 W/ft^2. Energy-efficient CFL (compact fluorescent lamps) with dimmable electronic ballasts have been used in most zones to achieve the LPD. Finally, solar heating is used for heating water, thereby conserving energy and utilizing plentiful renewable energy. The above measures have resulted in a 30% reduction in energy consumption.

CASE STUDY at a glance

- Wipro Technologies Ltd, one of the world's leading technology service providers, earned revenues in excess of Rs 100 billion in 2005/06.
- The company's Gurgaon Software Development Centre, located in Gurgaon, Haryana, has been designed and built on the 'green objectives' of energy conservation, water conservation, and waste management.
- Water efficiency is achieved through water-efficient landscaping, use of recycled greywater, and reduction of water usage through the installation of ultra low-flow water fixtures and rainwater harvesting systems.
- Resource conservation and measures to improve the built environment include the preservation of trees around the periphery of the site, the use of environment-friendly and recycled products such as aerated concrete blocks and recycled wood for construction, installation of double glazed windows, etc.
- Wipro conserves energy in its air-conditioning system by using variable frequency drives and motorized valves in all air-handling units, and through indirect evaporative cooling for precooling fresh air.
- Energy efficiency in the lighting system is achieved through efficient fixtures and controls, such as occupancy sensors, integration of daylight sensors, light shelves, and optimum glazing size.
- These efforts have led the Gurgaon Software Development Centre to achieve a Platinum-rating from the US Green Building Council in the Leadership in Energy and Environment Design category.
- Implementation of the above initiatives has been achieved with a nominal cost increase of 10% in construction, but has resulted in recurring energy savings of about 30%.

Annexe 1

About TERI–BCSD India

ENABLING RESPONSIBLE BUSINESS FOR A SUSTAINABLE FUTURE

Introduction

TERI–BCSD (Business Council for Sustainable Development) India is an independent and credible platform for corporate leaders to address issues related to sustainable development. It also seeks to promote leadership in achieving the triple bottom line, that is, environmental management, social responsibility, and economic performance. Initiated by TERI, the BCSD has now evolved into a strong industry body with membership coming from diverse sectors, including PSUs (public sector undertakings), MNCs (multinational corporations), and private sector companies from across India.

TERI–BCSD India is a partner of the WBCSD (World Business Council for Sustainable Development), Geneva, and a member of its regional network. The objective of TERI–BCSD India is to develop into a major forum for sustainable development in corporate India, represented by leaders of industry.

An industry-led structure

In order to make TERI–BCSD India an industry-led initiative, a member-driven executive committee, with CEOs (chief executive officers) from all sectors comprising its membership, has been set up. Two co-chairs and two vice-chairs elected from the membership lead the executive committee, which provides strategic direction and

TERI–BCSD India Executive Committee 2007

Co-chairs

Mr Thulasidas, Air India Ltd

Mr R S Sharma, ONGC Ltd

Vice-chairs

Mr K C Mehra, Forbes Gokak Ltd

Mr Ajay Shriram, DCM Shriram Consolidated Ltd

Members

Mr Ravi Sharma, Alcatel-Lucent

Mr Prasad Chandran, BASF India Ltd

Mr R R Vederah, Ballarpur Industries Ltd

Mr Mahesh Babu, IL&FS Ecosmart Ltd

Mr Manoj Gaur, Jaiprakash Associates Ltd

Mr Tantra Narayan Thakur, PTC India Ltd

Mr Tulsi R Tanti, Suzlon Energy Ltd

Mr Ravi Kant, TATA Motors Ltd*

Mr B Muthuraman, TATA Steel

*Representative to be nominated

formulates the plan of action of the business council. The executive committee not only represents, but also builds industry ownership in the activities of TERI–BCSD India.

TERI–BCSD India members work in unison with the researchers at TERI, and the structure of the business council reflects this partnership. TERI provides secretariat support to the business council and acts, as the permanent technical resource for the latter's activities. Dr R K Pachauri, Director-General of TERI, serves as the president of TERI–BCSD India.

Thematic advisory groups

Based on the interest of its members, TERI–BCSD India has organized itself into the following five TAGs (thematic advisory groups).

- Energy
- Water
- Climate change
- Health
- Sustainable livelihoods

Member companies nominate representatives for each group. Typically, members nominate representatives who possess expertise on each TAG theme, and who can identify issues on the themes concerning their CSR (corporate social responsibility) mandates.

The objective of each group is to conceptualize and implement activities and projects on the defined themes. The projects bring together the core competencies of member companies and the research and implementation strength of TERI (Table 1).

The Renewable Energy Chapter

The Renewable Energy Chapter of TERI–BCSD India has been conceived as a membership-based forum representing the renewable energy industry in India.

Its mission is to assist in creating an enabling environment for the growth of renewable energy in the Indian subcontinent. The common feature of the selected members will be their interest in meeting India's growing energy needs economically, through indigenous resources, and with the smallest environmental footprint.

Seminars and discussion forums

- **The CEO Forum**

  The annual CEO Forum is the curtain-raiser event of the DSDS (Delhi Sustainable Development Summit), and is a platform for global business leaders to develop and refine strategies for sustainable development.

- **Themes of discussion have included the following.**

  2005: CSR—Managing stakeholders' changing expectations
  2006: Business imperatives and opportunities—Water, energy, and climate change
  2007: Business and society—Partnering for a sustainable future

- **Talking tomorrow lecture series**

  The talking tomorrow lecture series features talks by eminent experts from across the globe, in the fields of energy, water, climate change, health, and sustainable development. Speakers offer valuable insight into these fields and their impact on the global business environment. Speakers at the talking tomorrow lecture series have included the following.

TABLE 1
Collaborations between TAGs and TERI departments

TAG	*Activities*
Health	**TERI–BCSD India** *Regular dialogue via the following* ▪ Video conference linking Indian companies in Chennai, Delhi, and Mumbai with the World Bank Institute, Washington, DC on the topic Corporate social responsibility and business competitiveness: The case for private sector workplace programmes to fight HIV/AIDS ▪ National Seminar for Indian industry: Business fight against HIV/AIDS: Towards a paradigm shift *Research* ▪ Modalities of implementing CSR (corporate social responsibility) and HIV/AIDS programmes in the supply chains of Indian companies Publication: Operationalizing a responsible supply chain: A primer for Indian companies
Energy	**Energy–Environment Technology Division** *Assessment of the energy efficiency of buildings in India, using TERI–GRIHA (Green Rating for Integrated Habitat Assessment)*
Water	Policy Analysis Division: Water Resources and Policy Management ▪ Training programme for corporate managers on water auditing techniques and tips on watershed modeling *Research* Water scenarios for Indian businesses: Exploring the key water challenges and drivers for Indian businesses by developing scenarios to 2025/50. BCSD India will leverage off the water scenario work already carried out by other organizations; in particular, the water scenarios of the WBCSD.
Climate change	**Policy Analysis Division: Centre for Global Environment Research** ▪ International seminar: Climate change-Redefining business imperatives and opportunities (November 2007). The seminar would provide a platform for the business and industry to discuss with the government and the scientific community, and explore a role for them to address climate change. Sustainable Development Outreach Division ▪ International Climate Change Exhibition (7–9 February 2008) Inspired by the theme of the 2008 edition of the annually held DSDS (Delhi Sustainable Development Summit), TERI will organize a climate change exhibition of international calibre. The exhibition aims to create an understanding between scientists, policy-makers, the broader public, and various other stakeholders on the dire climate change issues and measures to halt it.
Sustainable livelihoods	**Centre for IT (information technology) applications: IT and services** ▪ The rural ICT initiative-under which it will establish knowledge centres across some regions the country. These centres will seed the growth of social entrepreneur who will provide various services to the local village population. The knowledge centre will act as a 'single window' for the delivery of various services, products, and information.

TAG – thematic advisory group

November 2005
Dr Steve Koonin, Chief Scientist, BP Group
Talk on 'India's energy solutions: creating opportunities for growth in research and technology'

February 2006
Sir John Houghton, Trustee, Shell Foundation
Talk on 'Climate change: impact on the global business environment'

April 2006
Prof. Daniel C Esty, Director, Yale World Fellows
Programme and Hillhouse Professor of Environmental Law and Policy at Yale University
Talk on 'Topic benchmarking environmental sustainability with the 2006 Environmental Performance Index: why is India lagging?'

September 2006
Mr Herman Mulder, Former Senior Executive Vice President, ABN AMRO Bank NV and co-author of The Equator Principles
Talk on 'An emerging compelling business case: sustainable development'

Journals and publications

TERI–BCSD India publishes books and periodicals relevant to the corporate sector. A few publications are listed below.

- *Cleaner is cheaper*
- *Citizens at work* (documents the best CSR initiatives by the Indian corporate sector)
- *Business unusual* (highlights of the annual CEO Forum organized by TERI-BCSD India)
- CEO Alert (a bi-monthly newsletter for CEOs containing an analysis of sustainable development challenges relevant to the corporate sector)
- *EnCoRE* (a quarterly news magazine carrying updates on corporate environmental and social responsibility initiatives)
- TERI–BCSD India also brings out a weekly e-newsletter for its members called *TERI–BCSD clips*, which contains clippings of sustainable development news relevant to the business sector.

For further information please contact bcsd@teri.res.in
or visit the TERI–BCSD India website, <www.teriin.org/bcsd>.

Annexe 2

Contact details

Contact details of companies featured in this volume

Century Rayon, a division of Century Textiles and Industries Ltd
President
P B No. 22, Murbad Road
Shahad – 421 103

Chennai Petroleum Corporation Ltd
Director (Operations)
Manali
Chennai – 600 069

Grasim Industries Ltd, Staple Fibre Division
Sr. Executive President and Unit Head
Birlagram
Nagda – 456 331

Hindustan Zinc Ltd, Rampura Agucha Mine
Vice President
Rampura Agucha Mine
P O Agucha
Bhilwara, Rajasthan – 311 029

Indian Farmers Fertiliser Cooperative Ltd
Senior General Manager
P O IFFCO Township
Bareilly – 243 403

Indian Farmers Fertiliser Cooperative Ltd, Phulpur Unit
Sr. General Manager
P O Ghiya Nagar
Allahabad – 212 404

Indian Oil Corporation Ltd, Mathura Refinery
Chief Manager (S&EP)
Mathura Refinery
Mathura – 281 005

Indo Gulf Fertilisers, a unit of Aditya Birla Nuvo Ltd
Joint President (Manufacturing)
Jagdishpur Industrial Area
District Sultanpur – 227 817

Infosys Technologies Ltd
Associate Vice President(Facilities)
Plot No. 44 and 97A, Electronic city, Hosur Road
Bangalore – 560 100

JK Paper Mills, A Unit of JK Paper Ltd
Chief Executive (Works)
Jakaypur
Rayagada – 765 017

M/s Exide Industries Ltd
Chief Operating Manager
Survey No. 246, Chichurakanapalli
Sevaganapalli Panchayat, Hosur Taluk
Krishnagiri district
Tamil Nadu – 635 103

M/s Numaligarh Refinery Ltd
Dy. General Manager (Incharge, Refinery)
Numaligarh
Golaghat – 785 699

M/s Tata Sponge Iron Ltd
Asstt. General Manager (PS and Environment)
P O Joda
Keonjhar – 758 034

National Hydroelectric Power Corporation Ltd
Director (Personnel)
Sector-33,
Faridabad – 121 003

Oil and Natural Gas Corporation Ltd
ED-Chief HSE, ONGC
8th floor, Scope Minar
South Tower, Laxmi Nagar
New Delhi – 110 092

Oil and Natural Gas Corporation Limited, Cauvery Asset
GGM, Asset Manager
Cauvery Asset, Neravy Complex
Neravy
Karaikal – 609 604
Pondicherry

Oil and Natural Gas Corporation Ltd, Cauvery Asset
GGM-Asset Manager
Cauvery Asset, Neravy Complex
Neravy
Karaikal – 609 604
Pondicherry

Rashtriya Ispat Nigam Ltd
General Manager (Environment and Safety)
Room No. 303, EA (W) Building
Visakhapatnam – 530 031

Reliance Industries Ltd
Executive Director
Vill. Mora, P O Bhatha
Surat-Hazira Road
Surat – 394 510

Reliance Industries Ltd
President
B1-B5 MIDC Industrial Area
P O Patalganga
Raigad – 410 220

Sona Koyo Steering Systems Ltd
Asstt. General Manager, Environmental Management System
38/6, Delhi-Jaipur Road,
N H 8, P O Box No. 18
Gurgaon – 122 001

Toyota Kirloskar Motors Pvt. Ltd
Manager, Plant Administration Division
Regd. Office and works: Plot No.1, Bidadi Industrial Area
Ramnagar Taluk
Bangalore – 562 109

UAL-Bengal, UAL Industries Ltd
Vice President (Manufacturing)
16, Mayfair Road
Kolkata – 700 019

Vishay Components India Pvt. Ltd
Managing Director
Loni Kalbhor, C.Rly, Near Pune
Maharashtra – 412 201

Wipro Technologies Ltd
General Manager
No. 70/1,2,3,4, & 84/1,2,3,4
Keonics Electronic city, Hosur main road
Bangalore – 560 100